Water Treatment Operator Training Handbook

Water Treatment Operator Training Handbook

Fourth Edition

Nicholas G. Pizzi

American Water Works Association

Disclaimer
The authors, contributors, and publisher do not assume responsibility for the validity of the content or any consequences of their use. In no event will AWWA be liable for direct, indirect, special, incidental, or consequential damages arising out of the use of information presented in this book. In particular, AWWA will not be responsible for any costs, including, but not limited to, those incurred as a result of lost revenue. In no event shall AWWA's liability exceed the amount paid for the purchase of this book.

ISBN: 978-1-64717-247-3 ISBN, electronic: 978-1-61300-770-9
https://doi.org/10.12999/AWWA.20481-4E

Sr. Manager — Product Acquisition & Development: Geoffrey S. Shideler
Manager — Publishing Operations: Gillian Wink
Specialist — Copyright and Permissions: Peggy Tyler
Director of Publishing: John Fedor
Technical Editor: Wiley
Technical Editing Review: Suzanne Snyder
Cover Design and Technical Illustrations: Michael Labruyere
Cover Image: jasoncollinphotography.com
Production: Innodata

Library of Congress Cataloging-in-Publication Data
CIP data pending from the Library of Congress

American
Water Works
Association

Contents

List of Figures

List of Tables

Preface

Much has happened in the drinking water business since the last edition of this book was published in 2013. Some of it was highly publicized and reflected negatively on plant operators. In several cases, it was justified. When the operators of this country's water treatment facilities are found to have performed poorly or to have made bad and even criminal decisions, the general public takes notice. To that end, I am including a chapter on ethics and related types of training for plant operators.

What is also important to the public is the faith that the operators of their water systems possess the knowledge and skills to perform at a consistently high level, and that there is a steady flow of educable young staff willing to learn and take over the water systems. Therein lies the challenge.

The job duties and tasks of WTP operators have been upended by the improvements in supervisory control and data acquisition (SCADA) systems to a point that the daily operator routine of 25 years ago would be almost unrecognizable today. SCADA system design and maintenance has become such a specialized service that most utilities must rely on contractors to provide and maintain it. If the operator is unable to get into the system to perform simple tasks, such as formatting the y-axis on a trending graph or retrieving the data for analysis of filter behavior, the system is wasted.

In their laudable efforts to improve operator performance, designers have provided systems that can perform tasks much faster and efficiently. SCADA has the potential to improve plant operations to the benefit of all. However, many young operators are not learning to use as much of the basic knowledge that is needed to operate a plant manually. On those occasions when SCADA fails or is temporarily unusable, the staff may be unable to respond. In many plants today, we see that operators are not in control of their SCADA systems. They aren't in possession of their own data.

Other challenges, such as the need for enhanced cybersecurity protocols and the use of artificial intelligence in the business, take time from the operator that otherwise might be spent on learning how to treat water.

Cutbacks in state budgets have taken their toll on regulatory staffing capability. In many states, operator certification responsibilities have been abandoned by governmental agencies and contracted out to private companies who create exams that don't force the operator to learn as much about treatment as in the past. Primarily, new operators are expected to answer questions related to safety and management.

Plant managers have complained that they can no longer depend on the licensing process to help focus operator attention on the basics. They complain that while there is nothing wrong with learning safety and workplace human resources principles, those topics won't help much when the WTP is spinning out of control and critical response is needed.

The fourth edition of this book preserves the reservoir of knowledge found in the first three editions and blends in new material that is more challenging for the advanced operators who want to optimize their treatment processes. A chapter of practical problem-solving using a model plant has been added. It teaches operators how to work the math problems from actual plant data that are supplied. This edition also speaks to newly hired operators in an attempt to provide a traditional starting point for them that teaches them to think through problems and solve them on their own.

I especially want to thank my longtime friend and colleague Bill Lauer, who coauthored the third edition (2013) at a time when I was undergoing chemotherapy and could not finish that project. Bill kept this important resource alive and up to date with his insights and expertise. During our work together on the Partnership for Safe Water from 1997 to 2007, I learned to respect Bill's passion and work ethic.

ON THE PURPOSE AND LEGACY OF OPERATORS

From the beginnings of civilization, humans understood the necessity of living near an abundant supply of fresh water. They could not survive without it. Each family unit, each tribe, each culture needed an abundant supply of drinking water.

As the human population grew over the millennia, the overcrowding of some communities brought on unexplained illnesses thought to be caused by the "miasma"—polluted air reeking of the stench of human and animal wastes that accumulated in the villages and streets. It wasn't until the work of people like Dr. John Snow, who demonstrated that drinking polluted water could make you ill, and the subsequent discovery of bacteria as the causative agent of illnesses that the human population began to understand how precious the community water supply is to human health.

Soon, each town or village would come to depend on a steward of their local water supply to keep water fresh, abundant, and safe. Science provided the processes of filtration and disinfection, and the need for uniformity among those stewards became important. Thus was born the WTP operator—a person who sees that the community receives an abundant supply of safe water at all times. That person remains an important individual whose actions have great effect on the community they serve.

This book is dedicated to the operators of WTPs who toil, often anonymously, underpaid, and perhaps unappreciated, to serve up the best water possible.

Nick Pizzi

Chapter 1

Regulated Contaminants and Treatment Challenges

Operators learn early on that they must follow regulations or people can get sick or die. The communities that thrived over time understood that if their local water supply was treated to minimum standards, the health of the community would prosper. In time, it became logical to codify these standards into public law and provide enforcement of these minimum standards so that all consumers of water supplies could be assured that these minimum standards were being met. In the early 1900s, the US Public Health Service was formed, and it developed programs for "interstate water carriers" centered around coliform bacteria. Although the program was a step in the right direction, it was limited to towns that had a port or a railroad stop and was rarely enforced evenly from state to state.

By the 1960s, it was evident to scientists that our public water supplies were in need of greater protection. A system was developed that defined and standardized water quality across the nation and that provided science-based requirements that were enforceable. Thus, the US Environmental Protection Agency (USEPA) was born, and the Safe Drinking Water Act (SDWA) was passed in 1974 to protect public health.

Regulations that govern US water supply and treatment are developed by the USEPA under the SDWA. Most states administer USEPA regulations after adopting regulations that are no less stringent than federal rules, and, in some cases, states have adopted stricter regulations or have developed regulations for additional contaminants that are not regulated by the USEPA.

This chapter discusses current and anticipated USEPA regulations and the challenges that operators face in their efforts to comply with the regulations. Water system operators should consult their local and state regulatory agencies to verify applicable regulations that may be different than the federal regulations listed in this chapter. The chapter concludes with a discussion of selected contaminants that are commonly found in water, their significance, and the methods for their removal.

TYPES OF WATER SYSTEMS

The SDWA defines a public water system (PWS) as a supply of piped water for human consumption that has at least 15 service connections or regularly serves 25 or more individuals 60 or more days each year. By that definition, private homes, groups of homes with a single water source but with fewer than 25 residents, and summer camps with their own water source that operate less than 60 days per year are not PWS. However, they may be subject to state or local regulations, along with state and local well construction, and water quality requirements.

PWS are classified into three categories based on the type of customers served:

- Community water systems (CWS) serve a residential population of at least 25 people or 15 service connections year-round.
- Nontransient noncommunity water systems (NTNCWS) serve, on a regular basis, at least 25 of the same people for at least six months of the year (e.g., factories, schools).
- Transient noncommunity water systems (TNCWS) provide water to 25 or more people for at least 60 days of the year, but not to the same people and not on a regular basis (e.g., restaurant, campgrounds).

The rationale for these classifications is based on the differences in exposure to contaminants experienced by people using the water. Users of CWS and NTNCWS may be exposed to contaminants over an extended period of time, which would subject the consumer to both acute and chronic health effects. TNCWS serve short-term users, and exposure to contaminants would only be brief; therefore, there is no chronic exposure, but users would be subject to acute health effects. These distinctions, along with others such as population served and water source, are important because the primacy agency regulates these water systems differently. Acute contaminants will cause a serious negative health effect with a single exposure, e.g., microbial contaminants and nitrate. Chronic contaminants can cause health effects with long-term exposure, e.g., arsenic and lead (Pb). Any chronic contaminant may cause acute health effects if the concentration is extremely high.

Therefore, the monitoring requirements for both community and noncommunity water systems apply to all contaminants that are considered a health threat. CWS are required to monitor a comprehensive list of contaminants, including those that are microbial, chemical, and radiological. The monitoring of TNCWS and NTNCWS focuses on contaminants relevant to the specific type of noncommunity system and may be less frequent and less comprehensive than that for CWS.

Before examining the specific regulations that govern contaminants, the operator needs to know the difference among the three concepts used in the contaminant monitoring process: the maximum contaminant level goal (MCLG), the maximum contaminant level (MCL), and treatment technique (TT).

- The MCLG is set for most substances at a level at which there are no known, or anticipated, health effects. For those substances that are suspected carcinogens, the MCLG is set at zero.
- The MCL is set as close as feasible to the MCLG for substances regulated under the SDWA. The MCL is a level that is reasonably and economically achievable. This is the enforceable regulated level. Water systems that exceed an MCL must take steps to install treatment to reduce the contaminant concentration to less than the MCL.
- Where the USEPA has found it impractical to set an MCL, a TT has been established instead of an MCL. A TT is an enforceable procedure or level of technological performance that PWS must follow to ensure control of a contaminant.

With these concepts in mind, the various regulations can be examined. This discussion is not meant to be all-inclusive. Because the regulatory process is an ever-evolving one, the reader is cautioned that some of the stated facts presented in this discussion may have changed since the writing of this chapter. For up-to-date information, it is best to contact the local office of the regulatory authority in the district or state where the utility operates.

Table 1-1 contains just a few of the more common regulated contaminants and their respective MCL or TT. In many cases, tables in this chapter have been constructed by "cherry picking" data from the original USEPA tables, and they are provided for illustration purposes only. They are not intended to be used for regulatory purposes. For comprehensive views of the information in the Safe Drinking Water Act, operators should visit the USEPA website: www.epa.gov/sdwa.

Operations personnel are expected to know the regulatory limits for compounds encountered in their water supply. However, the number and

Table 1-1 Selected USEPA drinking water standards

Contaminant	MCL or TT *mg/L**
Total coliform	5% (monthly positives), if sampling 40 or more per month No more than 1 positive sample, if collecting less than 40 samples per month
Turbidity	0.3 ntu monthly or 1 ntu†
Chlorite	1.0
HAA5	0.060
TTHM	0.080
Chloramines (as Cl_2)	4.0
Chlorine (as Cl_2)	4.0
Chlorine dioxide (as ClO_2)	0.8
Arsenic	0.010

(continued)

Table 1-1 Selected USEPA drinking water standards (*Continued*)

Contaminant	MCL or TT *mg/L**
Copper	TT, action level = 1.3
Cyanide (as free cyanide)	0.2
Fluoride	4.0
Lead	TT, action level = 0.015
Mercury (inorganic)	0.002
Nitrate (measured as nitrogen)	10
Nitrite (measured as nitrogen)	1
Radium 226 and 228 (combined)	5 pCi/L
Uranium	30 µg/L

HAA5—sum of 5 haloacetic acids (mono-, di-, and trichloroacetic acids plus mono- and dibromoacetic acids), MCL—maximum contaminant level, TT—treatment technique, TTHM—total trihalomethane, USEPA—US Environmental Protection Agency

*The listed standards are numerical representations of the current USEPA drinking water standard and do not include the sample frequency or location and other important compliance information. For a complete definition of the standards, consult USEPA Drinking Water Standards.

†Turbidity less than or equal to 0.3 ntu for the combined filter effluent for 95% of the monthly samples. At no time can turbidity be greater than 1 ntu.

www.epa.gov/ground-water-and-drinking-water/national-primary-drinking-water-regulations

variety of regulated substances make it unlikely that operators would know all of the regulatory limits. Operators must rely on current references for the most accurate information. These are available from the regulatory agency responsible for the treatment plant.

DISINFECTION BYPRODUCT AND MICROBIAL REGULATIONS

Drinking water treatment, including use of chemical disinfectants such as chlorine, ozone (O_3), and chlorine dioxide, has been an important step in protecting drinking water consumers from exposure to harmful microbial contaminants. However, these chemical disinfectants can also react with organic and inorganic substances in the water to produce byproducts that may be harmful to drinking water consumers, particularly some susceptible segments of the population. Therefore, drinking water treatment using chemical disinfectants involves a delicate balancing act—i.e., adding enough disinfectant to control harmful microorganisms but not enough to produce unacceptably high levels of regulated disinfection byproducts (DBPs). The USEPA has enacted several regulations affecting microbial control and production of DBPs in groundwater and surface water supplies for small and large public drinking water systems. These rules are referred to collectively as the Microbial and Disinfection Byproduct Rules (MDBPs). MDBPs are a series of interrelated regulations that address risks from microbial pathogens and disinfectants/DBPs. Microbial protection for consumers of drinking water from public supplies is provided by provisions of current rules listed subsequently and are discussed in more detail later in this chapter:

- Filter Backwash Recycling Rule (FBRR)
- Ground Water Rule (GWR)
- Surface Water Treatment Rule (SWTR)
- Interim Enhanced Surface Water Treatment Rule (IESWTR)
- Long-Term 1 Surface Water Treatment Rule (LT1ESWTR)
- Long-Term 2 Surface Water Treatment Rule (LT2ESWTR)
- Stage 1 Disinfectants and Disinfection Byproducts Rule (Stage 1 DBPR)
- Stage 2 Disinfectants and Disinfection Byproducts Rule (Stage 2 DBPR)
- Revised Total Coliform Rule (RTCR)

Provisions of the DBPR are intended to protect drinking water consumers against the unintended public health consequences associated with consumption of treated drinking water containing residual disinfectants and DBPs produced from degradation of these residual disinfectants or reaction of disinfectants with organic and inorganic DBP precursors.

More details regarding the DBPR, including the current Stage 1 and Stage 2 DBPR, are described in this chapter. In the following discussion, the DBPR will be discussed first, followed by the microbial protection rules (i.e., SWTR, GWR, and RTCR).

DBPR

The Stage 1 and Stage 2 DBPR requirements discussed in the following sections focus first on two specific contaminants (i.e., total trihalomethane [TTHM] and the sum of 5 specific haloacetic acids [HAA5], which are mono-, di-, and trichloroacetic acids plus mono- and dibromoacetic acids) and then on other aspects of these regulations dealing with control or removal of DBP precursors ("enhanced coagulation"), bromate, chlorite, and residual disinfectants.

Stage 1 DBPR: HAA5 and TTHM Provisions

The Stage 1 DBPR was published in 1998 and established an MCL of 0.080 mg/L for TTHM (the sum of four trihalomethanes [THMs], which are chloroform, bromodichloromethane, dibromochloromethane, and bromoform) and 0.060 mg/L for HAA5. Although the MCLs for TTHM and HAA5 were officially written as 0.080 mg/L and 0.060 mg/L, respectively, the limits are commonly referred to as "80/60," or 80 μg/L and 60 μg/L. Although the numerical value of each MCL is an important consideration, an understanding of the methodology used to calculate the compliance value to compare it with this MCL is a subtle and equally important consideration in understanding compliance with the DBPR. Under the Stage 1 DBPR, the compliance value for TTHM and HAA5 was determined by calculating a running annual average (RAA) during the previous 12 months for each DBP for all monitoring locations within the system. Most systems were required to monitor quarterly (i.e., four times per year), although small groundwater systems (fewer than 10,000 individuals) may have been allowed to sample once a year. Typically, the RAA was based on four monitoring locations sampled quarterly, meaning that the RAA will be the average of 16 monitoring results each for HAA5 and TTHM.

Unlike acute toxicity risks, for which the exposure could be a single glass of water, cancer risks, such as those believed to be linked to TTHM and HAA5, involve longer periods of exposure (e.g., daily glasses of water spanning decades). For chronic exposures such as these, exposure to an excessively high concentration of a given cancer-causing agent will not necessarily result in the consumer getting cancer from this source. Conversely, a consumer exposed to a lower concentration every day for a lifetime could be more likely to develop cancer. Therefore, regulation of DBPs to reduce cancer risks is not based on limiting exposure to a single incident (i.e., not a "single hit") but, rather, aimed at reducing the repeated exposure over time.

Stage 2 DBPR: HAA5 and TTHM Provisions

The Stage 2 DBPR was published in 2006. This rule tightened requirements for DBPs, but compliance is not achieved by modifying the numerical value of the MCLs or by requiring monitoring of new constituents. Instead, the rule makes compliance more challenging than under the Stage 1 DBPR by (1) changing the way the compliance value is calculated and (2) changing the compliance monitoring locations to sites representative of the greatest potential for THM and HAA formation. These changes were made to ensure uniform compliance with the DBP standards across all areas of the distribution system—i.e., compliance is required at each sampling location.

The compliance value in the Stage 2 DBPR is called the locational running annual average (LRAA), which is calculated by separately averaging the four quarterly samples at each monitoring location. Compliance is based on the maximum LRAA value (see Table 1-2). Furthermore, the Stage 2 DBPR included several interim steps that led to the replacement of many existing Stage 1 DBPR monitoring locations, with new locations representative of the greatest potential for consumer exposure to high levels of TTHM and HAA5.

The long-term goal of the Stage 2 DBPR was to identify locations within the distribution system with the greatest potential for either TTHM or HAA5 formations and then base compliance on the LRAA at or less than 80/60 for each of these locations. Many of these locations were identified during the initial distribution system evaluation (IDSE).

The IDSE included monitoring, modeling, and/or other evaluations of drinking water distribution systems to identify locations representative of the greatest potential for consumer exposure to high levels of TTHM and HAA5. The goal of the IDSE was to evaluate several potential monitoring locations to justify selection of monitoring locations for long-term compliance with the Stage 2 DBPR.

Table 1-2 Example of LRAA calculations for Stage 2 DBPR

		Sampling Location µg/L			
Year	**Quarter**	**A**	**B**	**C**	**D**
1	3rd	52	68	63	66
1	4th	35	42	38	41
2	1st	47	49	42	43
2	2nd	18	42	45	37
LRAA		38	50	47	47
Maximum LRAA					**50**

DPBR—Disinfectants and Disinfection Byproduct Rules, LRAA—locational running annual average

One item to note regarding the Stage 2 DBPR as it applies to TTHM and HAA5 is that the goal was to find the locations in the distribution system where average annual levels of these DBPs are highest. TTHM formation increases as contact time (CT) with free or combined chlorine increases, although formation in presence of combined chlorine is limited. Therefore, establishing points in the distribution system with the highest potential for TTHM formation is related to points with maximum water age.

By contrast, peak locations for HAA5 are more complicated because microorganisms in biofilm attached to distribution system pipe surfaces can biodegrade HAA5. Consequently, increasing formation of HAA5 over time is offset by biodegradation, eventually reaching a point at which HAA5 levels decrease over time, even to the point at which they drop to zero. Figure 1-1 shows gradual reduction in HAA5 formation over time in a distribution system, followed by eventual decrease of HAA5 as water age increases (water age measured in tracer test). In chloramination systems, HAA5 formation is limited. In fact, ammonium chloride is added as a quenching agent in HAA5 compliance samples to halt HAA5 formation before analysis (*Standard Methods* 2023). Therefore, little additional HAA5 formation occurs after chloramination to offset HAA5 biodegradation occurring in the distribution system.

Stage 2 DBPR monitoring plan and compliance monitoring. Each water system must develop and implement a monitoring plan that details how it will meet all applicable requirements. The monitoring plan must include specific locations and schedules for collecting samples, along with how the water system will calculate compliance with MCLs, maximum residual disinfectant levels (MRDLs), and removal requirements. The compliance monitoring requirements vary by the population served, number of treatment plants or wells, source water type, the treatment used, and the disinfectant used.

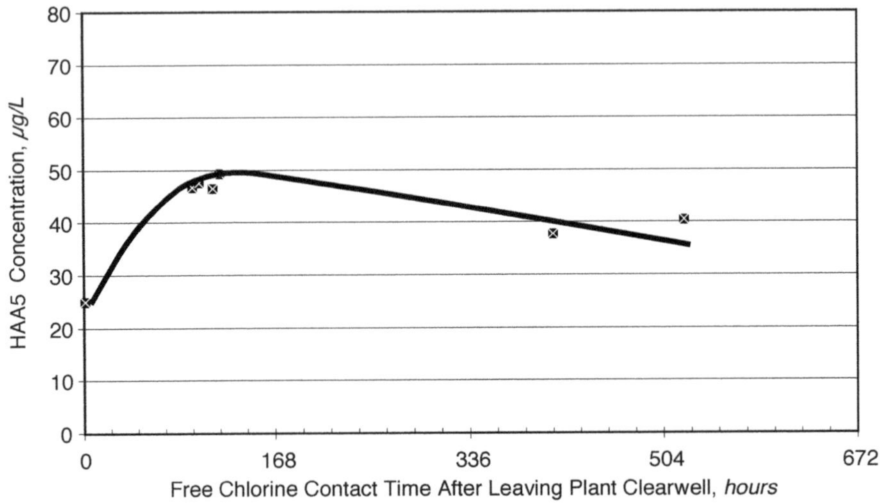

Figure 1-1 Formation and decay of HAA5 in a distribution system (time estimated by fluoride tracer test–T10)

Operational evaluation requirement of the stage 2 DBP. The Operational Evaluation (OEL) is intended for water systems to take proactive steps to assess source water, treatment, and distribution system processes to decrease high DBP levels and avoid MCL violations. Think of the OEL as the early warning system for when a system may exceed TTHM/HAA5 MCLs. The OEL is a projected average using TTHM and HAA5 results at each monitoring location and compared against the MCL: results for the two previous quarters plus twice the current quarter result divided by four. If the OEL is greater than the MCL for either TTHM or HAA5 at any monitoring location, it is an OEL exceedance. An OEL exceedance triggers the requirement that your system must perform an OEL and submit an OEL report to your state regulatory agency within 90 days. The evaluation must include an examination of system treatment and distribution practices, including storage tank operations, excess storage capacity, distribution flushing, changes in sources or source water quality, and treatment changes.

DBP Precursor Monitoring Requirements (TT)

Water systems with surface water or groundwater under the direct influence of surface water source that use conventional filtration must comply with the TT requirement and remove natural organic matter, measured as total organic carbon (TOC), that may react with disinfectants to form DBPs. Compliance with the TT can be achieved by removing specified percentages of TOC using enhanced coagulation or enhanced softening. The required TOC removal is listed in Table 1-3.

Table 1-3 TOC removal requirements

Source Water TOC *mg/L*	Source Water Alkalinity *mg/L as $CaCO_3$*		
	0–60	60–120	>120
>2.0–4.0	35.0%	25.0%	15.0%
>4.0–8.0	45.0%	35.0%	25.0%
>8.0	50.0%	40.0%	30.0%

www.epa.gov/sites/default/files/2020-06/documents/dbpr_plain_english_guide_final_508.pdf

$CaCO_3$—calcium carbonate, TOC—total organic carbon

You may also show that you are in compliance with the TOC removal requirement by using one of the following alternative compliance criteria:

1. RAA of monthly source water TOC samples less than 2.0 mg/L.
2. RAA of monthly treated water TOC samples less than 2.0 mg/L.
3. RAA of source water TOC less than 4.0 mg/L, RAA of source water alkalinity greater than 60 mg/L, and either of the following: (1) TTHM RAA less than or equal to 0.040 mg/L and HAA5 RAA less than or equal to 0.030 mg/L; or (2) The CWS or NTNCWS has made a "clear and irrevocable commitment" to installing technology to limit TTHM and HAA5 to those levels.
4. TTHM RAA less than or equal to 0.040 mg/L and HAA5 RAA less than or equal to 0.030 mg/L, and the CWS or NTNCWS uses only chlorine for primary disinfection and maintenance of a residual.
5. RAA of specific ultraviolet absorbance (SUVA) before any treatment less than or equal to 2.0 L/mg-m.
6. RAA of treated water SUVA less than or equal to 2.0 L/mg-m.
7. For softening systems only: softening that results in an RAA of treated water alkalinity less than 60 mg/L (as calcium carbonate).
8. For softening systems only: softening that results in an RAA for the removal of magnesium hardness (as calcium carbonate) of at least 10 mg/L.

NOTE: If a system cannot meet the required TOC removal levels or qualify through alternative compliance criteria, the system will need to do a bench- or pilot-scale testing to determine the coagulant dose and pH such that an incremental addition of 10 mg/L of alum or equivalent ferric salt results in a TOC removal of less than or equal to 0.3 mg/L. Results will need to be approved by the state regulatory agency.

Bromate

Bromate monitoring is only required for systems that use ozone for disinfection or oxidation. The MCL for bromate is 0.010 mg/L. Bromate can be present in systems using ozone where bromide is present at the ozone application point. Systems using ozone for oxidation or disinfection are required to monitor bromate levels once a month at the distribution system entry point, with compliance based on the RAA. Systems that use ozone and add sodium hypochlorite (NaOCl) need to closely monitor the quality of their sodium hypochlorite products for bromate content because bromate can potentially form during the manufacturing and storage of sodium hypochlorite.

Chlorite

Chlorite monitoring is required only for systems using chlorine dioxide as an oxidant or disinfectant because chlorite is a degradation product of chlorine dioxide. The MCL for chlorite is 1.0 mg/L. Chlorate, also a degradation product of chlorine dioxide, is not currently regulated. Monitoring chlorite is more complex than monitoring bromate because chlorine dioxide degrades over time, increasing chlorite formation. Chlorite monitoring requirements include daily monitoring at the distribution system entry point and monthly samples at three locations in the distribution system (i.e., first customer, average residence time, and maximum residence time). An exceedance at the entry point is not an MCL violation but triggers increased monitoring in the distribution system. If any daily entry point sample exceeds 1.0 mg/L, the system must take additional samples the following day at the three locations in the distribution system. These samples are collected on the same day. If the average chlorite value for the three follow-up samples exceeds 1.0 mg/L, the system violates the MCL. Compliance with the MCL for chlorite is based on the arithmetic average of each three-sample set. Although chlorite and chlorate can be degradation products of sodium hypochlorite, systems using sodium hypochlorite are not required to monitor for chlorite unless they also use chlorine dioxide.

Residual Disinfectants

The MRDL for combined or total chlorine is 4.0 mg/L as Cl_2. MRDL is based on the same data used to monitor minimum free and combined chlorine levels in the distribution system as required by the SWTR, using the same monitoring locations used for the RTCR. Compliance is based on running annual arithmetic average, computed quarterly, of monthly averages of all samples collected in the system. For systems switching between chlorine and chloramine during the year, compliance is determined based on results of all samples.

Chlorine dioxide residual has an MRDL of 0.8 mg/L as ClO_2, based on daily samples at the treatment plant. If daily samples exceed the MRDL (0.8 mg/L), the system must take follow-up samples in the distribution system the next day.

SWTRs

The SWTRs are developed with Stage 1 and Stage 2 DBPRs, are part of the suite of rules known as the MDBP rules, and are intended to reduce microbial contaminants in the water and, at the same time, minimize the risks posed by disinfectants and DBPs.

SWTRs Regulatory History

- SWTR, June 1989
- IESWTR, December 1998
- FBRR, June 2001
- LT1ESWTR, January 2002
- LT2ESWTR, January 2006

The goal of the SWTRs is to reduce illness caused by pathogens in drinking water, particularly *Cryptosporidium*, *Giardia*, and viruses. SWTRs apply to all PWS using surface water or groundwater sources under the direct influence of surface water (GWUDI).

Requirements:

- Set MCLGs for microbial contaminants.
- MCLGs of zero for *Cryptosporidium*, *Giardia*, viruses, and *Legionella*.
- Provide 99% (2-log) removal of *Cryptosporidium*, 99.9% (3-log) removal/inactivation of *Giardia*, and 99.99% (4-log) removal/inactivation of viruses.
- Provide 0.2 mg/L of disinfectant residual entering the distribution system and a detectable chlorine residual within the system.
- Combined filter effluent (CFE turbidity) requirements: CFE turbidity under 0.3 ntu in 95% of samples, and all measurements must be less than 1 ntu (compliance based on CFE samples collected at 4-hour intervals).
- Individual filter effluent (IFE) requirements: Continuously monitor IFE and recording at 15-min intervals and must report results, including a filter profile (graphical representation of filter performance), if either of the following two conditions are met: (1) turbidity in any filter for two consecutive 15-min intervals exceeds 1.0 ntu; or (2) turbidity during first 4 hours of a given filter run exceeds 0.5 ntu for two consecutive 15-min samples. Exceedances require follow-up action.
- Disinfection profiling and benchmarking requires systems to calculate levels of microbial inactivation to address risk tradeoffs of DBPs. The disinfection benchmark is used as a baseline of inactivation when considering and changes in the disinfection process. Systems must profile inactivation levels over 12 consecutive months. A disinfection benchmark is the lowest monthly average microbial inactivation value. States may waive the disinfection profiling requirement if TTHMs are less than 0.064 mg/L and HAA5s are less than 0.048 mg/L.
- Additional *Cryptosporidium* treatment requirements to higher risk systems based on source water monitoring and bin classification. Systems conducted the initial 24 months of source water monitoring for *Cryptosporidium. Escherichia coli* and turbidity levels to determine bin classification and sampling was repeated six years later to redetermine bin classification. Bin classifications determine whether any additional *Cryptosporidium* treatment is necessary. Facilities in the lowest bin (bin 1) require no additional treatment. Facilities in higher bins (bins 2–4) are required to either (1) provide additional *Cryptosporidium* protection from new facilities or programs not currently in use at a facility or (2) demonstrate greater *Cryptosporidium* protection capabilities of existing facilities and programs using a group of USEPA-approved treatment technologies, watershed programs, and demonstration studies, referred to collectively as the microbial toolbox.

Table 1-4 Inactivation requirements for *Giardia* and *Cryptosporidium*

Minimum Requirements for Inactivation or Removal of *Cryptosporidium*, *Giardia*, and Viruses								
Type of Filtration	Required Minimum Log Removal/Inactivation			Expected Log Removal by Filtration			Minimum Log Inactivation by Disinfection	
	Cryptosporidium (Log Removal Only)	*Giardia*	Viruses	*Cryptosporidium*	*Giardia*	Viruses	*Giardia*	Viruses
Conventional	2	3	4	2	2.5	2.0	0.5	2.0
Direct	2	3	4	2	2.0	1.0	1.0	3.0
Slow sand	2	3	4	2	2.0	2.0	1.0	2.0

Source: www.epa.gov/ground-water-and-drinking-water/national-primary-drinking-water-regulations

- Finished water reservoirs must be covered, or treat the discharge from the uncovered finished water storage facility to the distribution system to achieve at least 4-log virus, 3-log *Giardia*, and 2-log *Cryptosporidium* inactivation and/or removal using a state-approved protocol.
- Systems must conduct a sanitary survey every three years.
- Requires that watershed protection programs address *Cryptosporidium* for systems that are not required to provide filtration.
- Operated by qualified personnel as specified by the state regulatory agency.

Log Removal Credit

Facilities in compliance with SWTRs are designated to have provided 2-log virus removal, 2.5-log *Giardia* removal, and 2-log *Cryptosporidium* removal (Table 1-4). The additional credits (0.5-log for *Giardia* and 2-log for viruses) are required to be achieved by disinfection with either free chlorine, chloramines, ozone, or chlorine dioxide by meeting "CT" requirements described later in this chapter.

$C \times T$ Requirements and Log Inactivation

The SWTR (1989) requires all PWS using surface water or GWUDI sources to provide a 3-log or 99.9% *Giardia* removal/inactivation and a 4-log (99.99%) enteric virus removal/inactivation. The $C \times T$ value is used to determine log inactivation, which measures the strength of a disinfectant and the time needed to inactivate viruses and *Giardia* cysts. The USEPA defines $C \times T$ as the product of the disinfectant residual concentration (C) in milligrams per liter and the T in minutes. The equation for the value is $C \times T$ = mg/L × minute, which is typically expressed as "mg-min/L."

What is Log Inactivation?

"Log inactivation" is a convenient way to express the number or percent of microorganisms inactivated (killed or unable to replicate) through the disinfection process. For example, a 3-log inactivation value means that 99.9% of microorganisms of interest have been inactivated. Log inactivation measures the effectiveness of the disinfection process, which is influenced by variables including disinfectant concentration, temperature, pH, and disinfectant type (e.g., lower temperature results in less inactivation because the reactions slow down as temperature decreases).

What is C x T?

"$C \times T$" (minutes × milligrams per liter) in the context of water treatment is defined as the product of the following: *C*, for "residual disinfectant concentration" in milligrams per liter (determined before or at the first customer), and *T*, for the corresponding "disinfectant contact time" in minutes. $C \times T$ is a measure of the disinfection process reaction time.

Once the $C \times T$ value is calculated, it can be looked up in the USEPA CT tables to determine the log inactivation. The USEPA 3-log CT (99.9%) and 4-log $C \times T$ (99.99%) tables were developed by the USEPA, based on experimental data, to account for the effect of major variables (e.g., temperature, pH, concentration) on disinfection reaction efficiency. Water treatment plants (WTPs) are required to calculate log inactivation at least daily, and it must be higher than the minimum value specified by type of filtration, conventional, direct, or slow sand.

Sanitary surveys. Sanitary surveys are essential for ensuring safe drinking water, mandated under the IESWTR for systems with surface water or GWUDI. These onsite reviews evaluate the adequacy of the water source, facilities, equipment, and maintenance. They are proactive public health measures assessing a water system's ability to supply safe drinking water and offer an opportunity for education and technical assistance from SDWA primacy agencies.

Sanitary surveys typically cover the following eight fundamental areas:

1. Water sources
2. Water treatment processes
3. Water supply pumps and pumping facilities
4. Finished water storage facilities
5. Distribution systems
6. Monitoring, reporting, and data verification
7. Water system management and operations
8. Operator compliance with state requirements

Primacy agencies must conduct these surveys at least every three years for CWS and every five years for TNCWS and NTNCWS.

FBRR

FBRR applies to all PWS that use surface water or GWUDI; use direct or conventional filtration processes; and recycle spent filter backwash water, sludge thickener supernatant, or liquids from dewatering processes. The FBRR is intended to reduce the opportunity for recycle practices to adversely affect the performance of drinking water treatment plants (WTPs) and to help prevent microbes such as *Giardia* and *Cryptosporidium* from passing through treatment systems and into finished drinking water. Additionally, large volumes of recycled water may upset treatment processes, allowing contaminants to pass through the system. To minimize these risks, the FBRR TT requires that the recycle streams pass through all the processes of a system's existing conventional or direct filtration system. The FBRR also establishes reporting and recordkeeping requirements to assess the effect on overall treatment plant performance.

GWR

The GWR, published on Nov. 8, 2006, applies to all groundwater systems. The purpose of the GWR is to reduce the risk of illness caused by microbial contamination in public groundwater systems. The GWR established a risk-based approach to target groundwater systems vulnerable to fecal contamination.

The risk-based strategy addresses risks through an approach that relies on four major components:

- Routine sanitary surveys of systems that require the evaluation of eight critical elements of a PWS and the identification of significant deficiencies.
- Triggered source water monitoring for a system that identifies a positive sample from RTCR monitoring for systems that do not provide 4-log virus treatment.
- Corrective action is required for any system with a significant deficiency or source water fecal contamination, and corrective action can consist of correcting all significant deficiencies, providing an alternate source of water, eliminating the source of contamination, or providing treatment for 4-log inactivation of viruses.
- Compliance monitoring to ensure that treatment technology installed to treat drinking water reliably achieves 99.99% (4-log) inactivation or removal of viruses.

RTCR

The RTCR, published in 2013, updated the 1989 Total Coliform Rule. It is the only microbial drinking water regulation that applies to all PWS and mandates microbial testing based on the population served and system-specific sampling plans. The RTCR uses *E. coli* as a fecal indicator and sets an MCL for *E. coli*. The RTCR eliminated monthly public notification requirements based only on the presence of total coliforms. Total coliforms in the distribution system may indicate a potential pathway for contamination but, in and of themselves, do not indicate a health threat. Instead, the rule requires public notification when an *E. coli* MCL violation occurs, indicating a potential health threat, or when a PWS fails to conduct the required assessment and corrective action.

Routine sampling requirements:

- Routine monitoring is required based on system size and type.
- Collect total coliform samples from sites representing water quality throughout the distribution system, according to a written sample siting plan which is subject to state review and revision.
- Test each total coliform-positive (TC+) routine sample for *E. coli.*
- If any routine sample is TC+, repeat samples must be taken at the original sample tap within five service connections upstream and within five service connections downstream.
- Report any *E. coli*-positive sample to the state by the end of the day.
- Reduced monitoring may be available for PWS using only groundwater and serving fewer than 1,000 individuals.

Find-and-fix Approach: Assessment Triggers

The RTCR requires PWS to identify and correct sanitary defects and potential fecal contamination pathways through assessments based on TT triggers. Systems must address any sanitary defects within a specified time frame.

Level 1 assessment triggers:

- Failure to take every required repeat sample after a TC+ sample.
- PWS with fewer than 40 samples per month has two or more TC+ routine/repeat samples in the same month.
- PWS with at least 40 samples per month has more than 5.0% of routine/repeat samples in the same month that are TC+.

Level 2 assessment triggers:

- *E. coli* MCL violation.
- Second Level 1 assessment within a rolling 12-month time frame.
- Seasonal systems: A seasonal system is defined as a non-CWS that is not operated as a PWS on a year-round basis and starts up and shuts down at the beginning and end of each operating season. The seasonal system is required to have state-approved startup procedures, and the baseline routine monitoring frequency is monthly. Reduced monitoring may be available for seasonal PWS using only groundwater and serving fewer than 1,000 individuals.

Lead and Copper Rule and Revisions

(Note: the original language of the rule is saved here so that the history can be studied; the revisions follow.)

The objective of the Lead and Copper Rule (LCR) is to control corrosiveness of the finished water in drinking water distribution systems to limit the amount of lead (Pb) and copper (Cu) that may be leached from certain metal pipes and fittings in the distribution system. Of particular concern are pipes and fittings connecting the household tap to the distribution system service line at individual homes or businesses, especially because water can remain stagnant in these service lines for extended periods of time, increasing the potential to leach Pb, Cu, and other metals. Although the utility is not responsible for maintaining and/or replacing these household connections, they are responsible for controlling pH and corrosiveness of the water delivered to the consumers.

Details of the LCR include the following:

- The LCR became effective Dec. 7, 1992.
- The action level for Pb is 0.015 mg/L, and the action level for Cu is 1.3 mg/L.

A utility is in compliance at each sampling event (frequency discussed subsequently) when less than 10% of the distribution system samples are greater than the action levels for Pb and Cu (i.e., 90th percentile value for sampling event must be less than the action level).

Utilities found not to be in compliance must modify water treatment until they are in compliance. The term action level is used rather than MCL because noncompliance (i.e., exceeding an action level) triggers a need for modifications in treatment.

After identifying sampling locations and determining initial tap water Pb and Cu levels at each of these locations, utilities must also monitor other water quality parameters (WQPs) at these same locations as needed to monitor and evaluate corrosion-control characteristics of treated water. The only exemptions from analysis of these WQPs are systems serving fewer than 50,000 people for which Pb and Cu levels in initial samples are less than the action levels.

Pb, Cu, and WQPs are initially collected at six-month intervals, and then this frequency can be reduced if action levels are not exceeded and optimal water treatment is maintained. Systems that are in noncompliance and are performing additional corrosion-control activities must continue to monitor at six-month intervals, and they must collect WQPs from distribution system entry points every two weeks.

Each utility must complete a survey and evaluate materials that comprise their distribution system, in addition to using other available information, to target homes that are at high risk for Pb/Cu contamination.

Latest LCR revisions

On Jan. 15, 2021, the USEPA promulgated the LCR revisions (LCRR), and the deadline for water systems to comply with these revised requirements is Oct. 16, 2024. On Dec. 6, 2023, the USEPA published the proposed LCR improvements (LCRI) that, when final, will significantly reduce exposure to lead through drinking water. The LCRI proposal builds on the LCRR and the original LCR. The USEPA plans to finalize the LCRI before the LCRR compliance date of Oct. 16, 2024. For the LCRI, the USEPA proposed a compliance date of three years after the promulgation of the final rule and for systems to continue to comply with the LCR until that date, with the following exceptions. Specifically, the USEPA proposed to retain the 2021 LCRR Oct. 16, 2024, compliance date for the initial service line inventory, notification of service line material, Tier 1 public notification of a lead action level exceedance, and associated reporting requirements. With these limited exceptions, the USEPA proposed that water systems would directly transition from the LCR (as codified in the July 2020 *Code of Federal Regulations*) to the LCRI for all other rule provisions. As discussed in the USEPA's proposed rule for LCRI, transitioning from the LCR directly to the LCRI, as described earlier, would better enable systems and states to focus their resources on planning to comply with LCRI rather than short-term implementation of the LCRR. The USEPA anticipates that this focus would result in the earlier identification of more lead service lines, as well as the earlier replacement of full service lines, and that would more quickly reduce health risks associated with drinking water exposure to lead.

The USEPA finalized the LCRI in 2024. Goals of the LCRI include the following:

- Achieving 100% lead pipe replacement within 10 years.
- Locating legacy lead pipes. Knowing where lead pipes are is critical to replacing them efficiently and equitably. Water systems are currently required to provide an initial inventory of their lead service lines by Oct. 16, 2024. Under the proposed LCRI, all water systems would be required to regularly update their inventories, create a publicly available service line replacement plan, and identify the materials of all service lines of unknown material.
- Improve tap sampling, in which water systems would be required to collect first-liter and fifth-liter samples at sites with lead service lines and use the higher of the two values when determining compliance with the rule.
- Lowering the lead action level from 15 to 10 ppb. When a water system's lead sampling exceeds the action level, the system would be required to inform the public and take action to reduce lead exposure while concurrently working to replace all lead pipes.

Chemical Contaminant Rules

The Phase I, II, and V regulations, or the Chemical Contaminant Rules, were finalized in 1989, 1992, and 1995, respectively. These rules regulate over 65 contaminants in three contaminant groups: inorganic contaminants, including nitrate and arsenic; volatile organic contaminants; and synthetic organic contaminants (SOCs). The rules set MCLGs, MCLs, and monitoring requirements. MCLs and monitoring frequency vary with source type (groundwater or surface water), contaminant, PWS type, and population served. Monitoring frequency can be reduced or monitoring can be waived depending on the contaminant, vulnerability assessment, and previous analytical results. Nitrate, an acute contaminant, cannot be reduced to less than annually and no waivers are allowed, while SOCs, a chronic contaminant, can be reduced or waived.

Public Notification Rule (Right-to-Know Rule)

The Public Notification Rule (PNR) was revised in 2000, and it requires PWS to inform their customers about water quality issues, including violations of drinking water standards and potential health risks. Notices must be provided to individuals served, not just billing customers. Notifications must be timely and, clear and provide guidance on any necessary actions that customers should take. The rule specifies distribution methods (e.g., television, radio, newspaper, hand-delivery, mail), notice format, specific wording, and inclusion of information in languages other than English.

There are three tiers of notification based on the severity and type of violation, and the clock for notification starts when the PWS learns of the violation:

- Tier 1: Situation with potential to immediately affect human health (acute risk). PWS has hours to notify the public. The state regulatory agency has discretion to require Tier 1 for "other violations and situations."
- Tier 2: No immediate risk to public health, contaminant levels exceed MCLs, or failure to provide adequate treatment. PWS notifies the public within 30 days.
- Tier 3: No direct effect on public health—e.g., monitoring and reporting violations. Notify within one year or in the consumer confidence report (CCR).

CCR (Right-to-Know Rule)

The CCR rule mandates CWS to provide annual drinking water quality reports by July 1, detailing water sources, contaminant risks, and system compliance.

On May 15, 2024, the USEPA announced revisions to make these reports clearer and more accessible by 2027, including support for translations and better information on lead. The updated rule encourages electronic distribution and requires systems serving over 10,000 customers to provide reports twice a year. States must also submit annual compliance data to the USEPA.

Unregulated Contaminant Monitoring Rule

The USEPA's Unregulated Contaminant Monitoring Rule (UCMR), established under the SDWA Amendments of 1996 and the America's Water Infrastructure Act of 2018, aims to monitor unregulated contaminants in drinking water. The USEPA uses the UCMR program to collect nationally representative data for contaminants that may be present in drinking water but are not currently subject to regulatory standards set under SDWA. This monitoring is used by the agency to understand the frequency and level of occurrence of unregulated contaminants in the nation's PWS. Every five years, taking into consideration the USEPA's Contaminant Candidate List, the agency develops a new list of UCMR contaminants for monitoring. The USEPA analyzes the data to determine whether new drinking water regulations are needed. The UCMR is repeated every five years, with each cycle requiring utilities to sample a new list of contaminants. Analytical results are available in a National Contaminant Occurrence Database for drinking water. State and local officials may also use the UCMR data to assess the need for actions to protect public health.

- UCMR 1 (2001–2005)
- UCMR 2 (2007–2011)
- UCMR 3 (2012–2016)
- UCMR 4 (2017–2021)
- UCMR 5 (2022–2026)

Monitoring requirements:

- Large PWS (serving more than 10,000 people) must monitor all listed contaminants.
- Small PWS (serving 3,300–10,000 people) are included on a statistically random basis; the USEPA pays for the analysis of UCMR samples from participating PWS with fewer than 10,000 people.
- Reporting: Water systems subject to UCMR must comply with the CCR and PNRs.

Reporting requirements:

- The CCR rule requires CWS to report monitoring results when unregulated contaminants are measured at or greater than the minimum reporting level, including the average and range of the year's results.
- The PNR requires PWS to notify customers about UCMR results within 12 months, which can be fulfilled through the annual CCR delivered to all billing customers by July 1.

Compliance:

- Noncompliance includes failing to perform required sampling and analysis, use the appropriate analytical procedures, or report results are violations of the UCMR.
- However, the numerical results of these analyses cannot result in a violation because none of the constituents in the UCMR are currently regulated (i.e., no MCLs, action levels, or other standards apply).

UCMR 5 (2023–2026)

UCMR 5 was published on Dec. 27, 2021. UCMR 5 required sample collection for 30 chemical contaminants between 2023 and 2025 using analytical methods developed by the USEPA and consensus organizations. This action provided the agency and other interested parties with scientifically valid data on the national occurrence of these contaminants in drinking water.

Consistent with the USEPA's PFAS Strategic Roadmap, UCMR 5 provided new data that improve the agency's understanding of the frequency that 29 per- and polyfluoroalkyl substances (PFAS) and lithium are found in the nation's drinking water systems and at what levels. The monitoring data on PFAS and lithium help the USEPA make determinations regarding future regulations and other actions to protect public health under SDWA. The data also ensure science-based decision-making, help the agency better understand whether these contaminants in drinking water disproportionally affect communities with environmental justice concerns, and allow the USEPA, states, tribes, and PWS to target solutions.

Operator Certification

Amendments to the 1996 SDWA required the USEPA to develop national guidance for operator certification. The final rule was published on Feb. 5, 1999, and became effective on Feb. 5, 2001. State operator certification programs were required to address nine baseline standards, including operator qualifications, certification renewal, and program review. The USEPA will withhold 20% of a state's Drinking Water State Revolving Fund capitalization grant funds if the state has not adopted and implemented a compliant operator certification program or submitted an equivalent existing program. Although specific requirements vary by state, the ultimate goal of state programs is to ensure the treatment and distribution of safe drinking water.

Arsenic

The MCL for arsenic was reduced from 50 to 10 μg/L in the *Federal Register* published on Jan. 22, 2001. Samples are collected at the entry point to the distribution system. Surface water systems monitor annually, and groundwater systems monitor every three years. If results are greater than the MCL, the monitoring frequency may increase to quarterly sampling.

Radionuclides Rule

The Radionuclides Rule, published in December 2000, establishes monitoring requirements for CWS for radium-226, radium-228, gross alpha particle activity, and beta particle and photon radioactivity. States may grant waivers or reduced monitoring based on historical data and vulnerability assessments.

PFAS Regulation

PFAS are a series of synthetic organic chemical compounds that persist in the environment for lengthy periods of time. They are often called "forever chemicals." These PFAS chemicals have been used in industry and consumer products such as nonstick cookware, waterproof clothing, and stain-resistant furniture and have been important for certain industries and uses. Science shows that these chemicals are harmful to human health. PFAS exposure over an extended time can cause cancer and other serious illnesses that decrease quality of life or result in death. PFAS exposure during critical life stages such as pregnancy or early childhood can also result in adverse health effects.

On Apr. 26, 2024, the USEPA published the final National Primary Drinking Water Regulations (NPDWR) for six PFAS, establishing legally enforceable levels—i.e., MCLs—for six PFAS in drinking water. The USEPA also finalized health-based, non-enforceable MCLGs for these PFAS.

The six specific PFAS compounds included in the final regulation are as follows:

- Perfluorooctanoic acid
- Perfluorooctane sulfonic acid
- Perfluorononanoic acid
- Hexafluoropropylene oxide dimer acid, also known as GenX
- Perfluorohexane sulfonic acid
- Perfluorobutane sulfonic acid

The MCLs and MCLGs for these chemicals are shown in Table 1-5.

Table 1-5 PFAS MCLGs and MCLs

Chemical	MCLG	MCL
PFOA	0	4.0 ppt
PFOS	0	4.0 ppt
PFNA	10 ppt	10 ppt
PFHxS	10 ppt	10 ppt
HFPO-DA	10 ppt	10 ppt
Mixture of two or more: PFNA, PFHxS, HFPO-DA, and PFBS	HI of 1	HI of 1

HFPO-DA—hexafluoropropylene oxide dimer acid, HI—hazard index, MCL—maximum contaminant level, MCLG—maximum contaminant level goal, PFAS—per- and polyfluoroalkyl substances, PFBS—perfluorobutane sulfonic acid, PFHxS—perfluorohexane sulfonic acid, PFOA—perfluorooctanoic acid , PFOS—perfluorooctane sulfonic acid, PFNA—perfluorononanoic acid, USEPA—US Environmental Protection Agency

MCLG: The level of a contaminant in drinking water below which there is no known or expected risk to health. MCLGs allow for a margin of safety and are non-enforceable public health goals.

MCL: The highest level of a contaminant that is allowed in drinking water. MCLs are set as close to MCLGs as feasible using the best available treatment technology and taking cost into consideration. MCLs are enforceable standards.

HI: The HI is an approach that the USEPA regularly uses to understand health risk from a chemical mixture (i.e., exposure to multiple chemicals). The HI is made up of a sum of fractions. Each fraction compares the level of each PFAS measured in the water with the health-based water concentration.

Timeline

At the time of writing, PWS will have three years to comply with monitoring requirements. Within three years of rule promulgation (2024 to 2027), the following must happen:

- Initial monitoring must be complete.
- Initial monitoring of four quarterly samples for larger systems serving populations over 10,000 people and biannual samples from smaller systems serving populations of 10,000 or fewer.

Starting three years after rule promulgation (2027–2029), the following must happen:

- Results of initial monitoring must be included in CCRs (i.e., Annual Water Quality Report).
- Regular monitoring for compliance must begin, and results of compliance monitoring must be included in CCRs.
- Public notification for monitoring and testing violations must begin.

Starting five years after rule promulgation (starting 2029), the following must happen:

- Take steps to reduce level of PFAS and comply with all MCLs.
- Public notification for MCL violations must begin.

Analytical Methods

Each of the individual USEPA regulations contains their own information regarding analytical methods approved for compliance monitoring. These and other approved analytical methods are compiled in a final rule titled "Analytical Methods for Chemical and Microbiological Contaminants and Revisions to Laboratory Certification Requirements" published Dec. 1, 1999. These analytical methods were approved for compliance monitoring effective Jan. 3, 2000. The USEPA-approved methods include analytical procedures developed by the USEPA, plus procedures developed by others that the USEPA endorses, including specific procedures developed by ASTM International and some specific procedures included in *Standard Methods* (2023).

Currently, only approved analytical methods can be used for compliance monitoring. In the future, the USEPA hopes to implement a performance-based measurement system that will allow utilities to use alternative screening methods instead of requiring only USEPA-approved reference methods. The 1996 SDWA Amendments require the USEPA to review new analytical methods that may be used for the screening and analysis of regulated contaminants. After this review, the USEPA may approve methods that may be more accurate or cost-effective than established methods for compliance monitoring. These screening methods are expected to provide flexibility in compliance monitoring and may be better and/or faster than existing analytical methods.

The approval of new drinking water analytical methods can be announced through an expedited process in the *Federal Register*. This allows laboratories and water systems more timely access to new alternative testing methods than the traditional rulemaking process. If alternate test procedures have the same or better performance of the approved methods, they can be considered for approval using the expedited process.

Operational Views of Certain Contaminants

Turbidity

Turbidity is the measure of the amount of particulate material in water. It is measured by detecting the amount of light scattered by the particles in a water sample. Turbidity is used as an indicator of water quality and as an indicator of the efficiency of certain removal processes such as coagulation and filtration. Also, it is easily measured in the lab and, therefore, is perfect for goal-setting processes. Adequate removal of turbidity is an important step in the process of removing pathogens. Although the turbidity measurement provides no information regarding the nature of the particles it measures, turbidity is considered very important and is thus regulated as a TT. Optimization efforts, such as those found in the Partnership for Safe Water, are designed around the optimization of turbidity removal. High turbidity levels can make the disinfection process less efficient by creating higher disinfectant demand. Higher turbidities may also protect coliforms from disinfection by absorbing or encasing them.

Turbidity acts as a *surrogate* for pathogens. Pathogen passage can be related to turbidity events (spiking) in the finished water. The minimization of the frequency, magnitude, and duration of spiking should be a top priority in water treatment. Operators should view any event of individual filter turbidity spiking as a potential for the passage of harmful pathogens into the water column. Each event should be analyzed for cause, and a plan for the elimination of the cause(s) should be implemented. Most filter-spiking events can be traced to operator involvement, unless plant inadequacies exist that make it impossible for water to be treated at times.

TOC

TOC is a composite measure of the organic content of the water. It is important for water suppliers to measure this contaminant because its presence and amount correlate to the production of DBPs. TOC can be removed in the coagulation/settling/filtration process, and removal efficiencies for this contaminant are regulated.

The challenge for treatment plant operators is to strike a balance between removal of TOC with traditional pH-lowering methods (more coagulant, more acid) and the need to maintain compliance levels with the LCR. Also, site-specific requirements for residuals disposal can be affected by the use of more coagulant for TOC removal. Especially difficult are the choices that operators consider when, although they are in compliance with DBP MCLs, they know they can achieve better results with more TOC removal through addition of more coagulant.

Waterborne Pathogens

Disease-causing organisms, called pathogens, include all of the problem-causing bacteria, viruses, protozoa, and algae. Well-known diseases caused by these organisms include typhoid fever, cholera, Legionnaires' disease, peptic ulcers, hepatitis, giardiasis, cryptosporidiosis, gastroenteritis, etc. All of these pathogens can be found in natural surface water supplies and, therefore, can invade the treated water supply in sufficient numbers that will cause disease in the consumer. Even groundwater supplies may come under the influence of surface water and, therefore, be contaminated. Typically, pathogen exposure presents an acute risk rather than a chronic risk, and the detected presence of them or their surrogates will bring on boil-water notices. Given that some consumers are temporarily or permanently less immune than other consumers, pathogen removal is among the most important tasks confronting the WTP operator.

Bacteria, which are organisms that can cause severe illness in people, are usually minimized by conventional water treatment processes. They can exist in wide variations of temperature and water quality ranges, and some bacteria can pass through the WTP and find their way to the consumer in small numbers. *Legionella* is such an example of an organism that can occur in the finished waters of systems using full treatment. Of particular interest to WTP operators is the group of bacteria called coliforms. These organisms are used as indicator bacteria because they are easily cultured in the lab and can be an indicator of the presence of other bacteria that are more difficult to culture.

Viruses, which are smaller organisms than bacteria, can also cause many debilitating diseases, including hepatitis and polio. They are difficult to culture, even in the most sophisticated labs. Fortunately, most viruses are removed or inactivated by conventional water treatment practices. The lime-softening process is very effective against viruses because of the high pH used, and maintaining proper CT values improves disinfection of viruses.

Protozoa are larger than bacteria and can be very resistant to disinfection because of their ability to form spores or cysts that have tough outer casings. The notable protozoa that WTP operators deal with are *Giardia* and *Cryptosporidium*. These two organisms are found in nature and are typically removed in the filtration process. *Giardia* can be inactivated by sufficient $C \times T$ values using chlorine, but *Cryptosporidium* is considered resistant to chlorine at levels far greater than those typically used in WTPs. The IESWTR regulates these two organisms (see related text in this chapter) by requiring TTs. When the TTs are met, the utility is given "credit" for removal of *Cryptosporidium* and removal/inactivation of the *Giardia*. This credit must be achieved continually as the plant operates. Failure to meet the treatment technique results in boil-water notices.

Algae do not typically pose a health concern for humans, although some may produce neurotoxins and hepatotoxins such as microcystin that can be very harmful to humans and pets. The more frequent problems caused by algae are taste-and-odor episodes and filter-clogging episodes. Algae are difficult to coagulate and filter, and it is common for them to pass through filters in high numbers. Because of this, it is often more practical to prevent algae from growing than it is to remove them once they are in the plant.

Inorganic Chemical Contaminants

The inorganic chemical contaminants found in water supplies may be found as naturally occurring in the source, or they may appear in the water through contact with the piping and storage components of the system. The chemical treatment process can also add to inorganic contamination. These chemicals in water can present chronic health problems at low levels and can also cause acute distress when ingested in high doses.

Many inorganic chemical contaminants have MCLs associated with them. Notable exceptions are lead and copper, which carry action levels, as explained previously in the chapter. Most are regulated at the entry point to the distribution system, again with lead and copper being the exception. Iron and manganese are of special interest to the operator because of the staining problems that they can create on fixtures. Removal of these nuisance contaminants is detailed in Chapters 4 and 10. Nitrates are regulated with an MCL of 10 mg/L because they are believed to cause problems with the circulatory systems of infants.

Fluoride (see Chapter 10) is an inorganic constituent with an interesting role in the water treatment business. Its presence in water is required by most states and communities (there are exceptions), but it is also a regulated contaminant. Adding to the confusion is the fact that fluoride carries with it both an MCL and a secondary MCL (SMCL). The WTP operator, therefore, is confronted with the problem of adding a regulated contaminant to the water and then carefully controlling the amount that is added. The MCL for fluoride is 4.0 mg/L, and the SMCL is 2.0 mg/L. High levels of fluoride can cause mottling of the enamel of the teeth. Still-higher levels can be poisonous to humans.

DBPs

Chlorine has played the primary role in disinfection of PWS, but chlorine dioxide, ozone, and chloramines are also used. The use of these disinfectants can cause the formation of DBPs, as previously discussed. Some of these byproducts are probably carcinogens and may cause other toxic effects. The rules governing the limitation of the DBPs set limits based on chronic exposure and tend to be regulated as averages for occurrence. The challenge for operators is to maintain adequate disinfection levels for microbiological protection while providing a water that is low in the byproducts created in the disinfection process.

Chief among the DBPs (where chlorination is used) is the occurrence of THMs (see Chapter 8), but the HAAs are also currently regulated. In time, other DBPs are expected to be regulated as health-effects data are brought forth.

When ozone is used as a primary disinfectant, there are potential concerns with bromate and formaldehyde production.

PFAS Treatment

PFAS are highly soluble in aquatic environments and can dissolve in surface water and groundwater. Because of their chemical and biological stability, PFAS are difficult to degrade via biodegradation, photolysis, or hydrolysis.

At the time of this book revision (summer to fall 2024), activated carbon, ion exchange resin, and membrane filtration treatment processes were being used to treat PFAS in drinking water, and early data show that treatment performance varies based on the compound and site.

Regulatory Changes

There is no doubt that regulations will continue to change based on results of research and a better understanding of the health consequences of various contaminants. The 1996 Safe Drinking Water Amendments require the USEPA to review existing NPDWRs at least every six years and revise, if appropriate. A determination to potentially revise regulations initiates a process that will involve more detailed analyses of health effects, analytical and treatment feasibility, occurrence, benefits, costs, and other regulatory matters. The USEPA announced the review results for the third Six-Year Review (called Six-Year Review 3). Based on the agency's detailed review of 76 NPDWRs, the USEPA concluded that 8 NPDWRs are candidates for regulatory revision. The eight candidates are chlorite, *Cryptosporidium*, HAAs, heterotrophic bacteria, *Giardia lamblia*, *Legionella*, TTHMs, and viruses.

The USEPA is currently conducting an analysis to evaluate the potential revision of MDBPs. The USEPA is expected to propose revisions in summer 2025 and to finalize these standards by summer of 2026.

WTP operators must keep abreast of the latest developments and be prepared to adjust to these regulations. Water system operators will occasionally be presented with a dilemma when new research indicates that an unregulated contaminant may be harmful and is found (perhaps in a very low concentration) in their water supply. What should the water utility tell its customers about this situation? Each utility should develop a communications plan and have it ready for this almost inevitable circumstance.

BIBLIOGRAPHY

AWWA. 2011 (6th ed.). *Water Quality and Treatment* (J.K. Edzwald, editor). New York: McGraw-Hill.

AWWA. 2010 (4th ed.). *Principles and Practices of Water Supply Operations—Water Treatment*. Denver: AWWA.

Brown, R.A. 2004. Personal communication.

Pizzi, N.G. and M. Rodgers. 2000. "Testing Your Enhanced Coagulation Endpoint." *Opflow*, 26(2):1-5.

Roland, M. 2024. Personal communication.

Standard Methods for the Examination of Water and Wastewater. 2023 (24th ed.). Washington, D.C.: APHA, AWWA, and WEF.

USEPA (US Environmental Protection Agency). n.d. "Ground Water and Drinking Water." http://water.epa.gov/drink/index.cfm (accessed August 25, 2025).

USEPA. n.d. "Ground Water Rule." www.epa.gov/dwreginfo/ground-water-rule (accessed August 25, 2025).

USEPA. n.d. "Revised Total Coliform Rule and Total Coliform Rule." www.epa.gov/dwreginfo/revised-total-coliform-rule-and-total-coliform-rule (accessed August 25, 2025).

USEPA. 2003. *Final Implementation Guidance for the Stage 1 Disinfectants/Disinfection Byproducts Rule*. LLLL: PPPP. http://water.epa.gov/lawsregs/rulesregs/sdwa/stage1/upload/s1dbprimplguid.pdf (accessed August 25, 2025).

USEPA. 2024. *2021 LCRR Requirements That Are Retained in the Final Lead and Copper Rule Improvements and Compliance Dates*. EPA 816-F-24-018. www.epa.gov/system/files/documents/2024-04/revised-508_lcrr-compliance-fact-sheet_4.17.24.pdf (accessed August 25, 2025).

USEPA. 2024. *Final PFAS National Primary Drinking Water Regulation*. www.epa.gov/system/files/documents/2024-04/pfas-npdwr_fact-sheet_general_4.9.24v1.pdf (accessed August 25, 2025).

USEPA. 2013. *Guidance Manual for Conducting Sanitary Surveys of Public Water Systems; Surface Water and Ground Water Under the Direct Influence (GWUDI)*. Washington USEPA. June 22, 2013.

USEPA. 1993. "Appendix E: Inactivation Achieved by Various Disinfectants." In *Guidance Manual for Compliance With the Filtration and Disinfection Requirements for Public Water Systems Using Surface Water Sources*. Washington: Science and Technology Branch, Criteria and Standards Division, Office of Drinking Water, USEPA.

Chapter 2

Basic Properties of Water

The earth's water is ancient. All that is on the surface of the earth, and all that is hidden below ground, has existed since the beginning of time. This water moves in a constant cycle from the surface of the earth to the atmosphere by the processes of evaporation and transpiration and moves back to the earth by the processes of condensation and precipitation. Some of the water that falls back to earth percolates into the soil and becomes groundwater, while other water stays on the surface in lakes and streams. This constant recycling of water is called *the hydrologic cycle* (Figure 2-1). This cycle is important to water treatment plant (WTP) operators because the sources of supply that they draw from can only be replenished by the hydrologic cycle. Knowledge of this replenishment process provides a sense of how fragile the cycle is and how the need for conservation and stewardship of this precious resource is paramount.

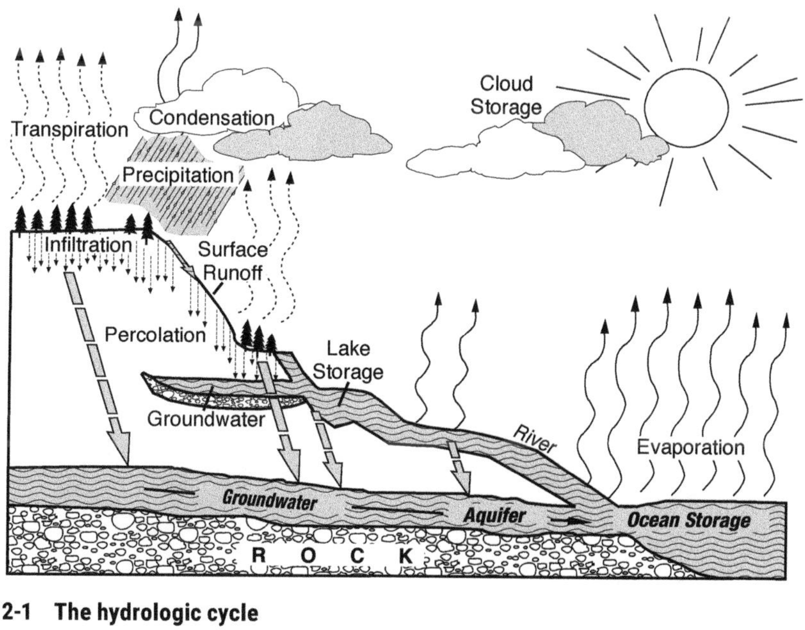

Figure 2-1 The hydrologic cycle

Water chemistry is not the purpose of this book, but operators should know some of the basic chemical and physical properties of water. Much can be learned by starting with the water molecule itself. The combination of the elements hydrogen (H) and oxygen (O) creates a unique compound that we know as the water molecule, and its properties are well known. WTP operators exploit these properties in their water treatment tasks.

The water molecule is made of two units of H and one unit of O, so chemists use shorthand when they refer to water as H_2O. They also tell us that each H has a slight positive electrical charge, while the O has an overall –2 charge. So together, the molecule is neutral in charge because the two negatives come together with the positive +2 charge for an overall charge of zero (Figure 2-2).

But the water molecule is not that simple in its nature. This miraculous molecule that does so much for us and for the world contains what from a distance seems to be a simple structure and purpose but, on closer inspection, reveals its secret: the hydrogen bond.

Although the water molecule has an overall neutral or "zero" electrical charge, it is polar in nature, meaning that it has a relatively negatively charged area near the center of the nucleus and a somewhat positively charged area near its outer boundaries, where the hydrogen atoms are located. Because of this arrangement, it has polarity—it is dipolar. That means it will attract other molecules that are dissolved in the water, and it will help dissolve still others. Those attributes affect treatment in many ways.

Because of this, water is great at dissolving many substances it comes into contact with, so it is not likely that we will find it pure in nature. Rather,

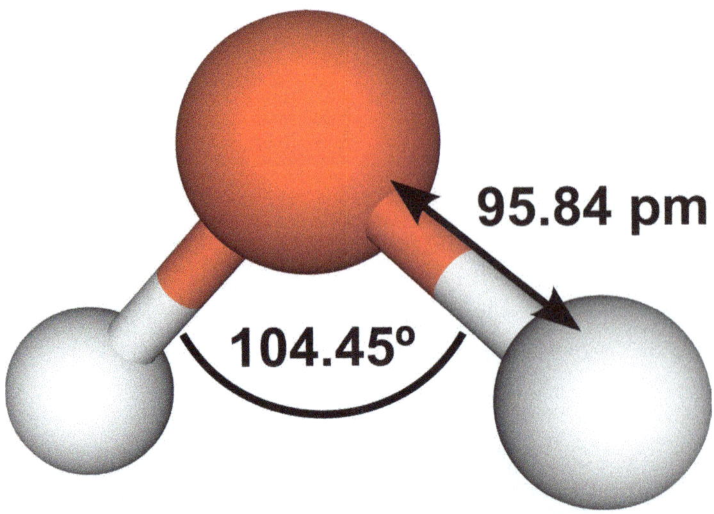

Figure 2-2 The water molecule

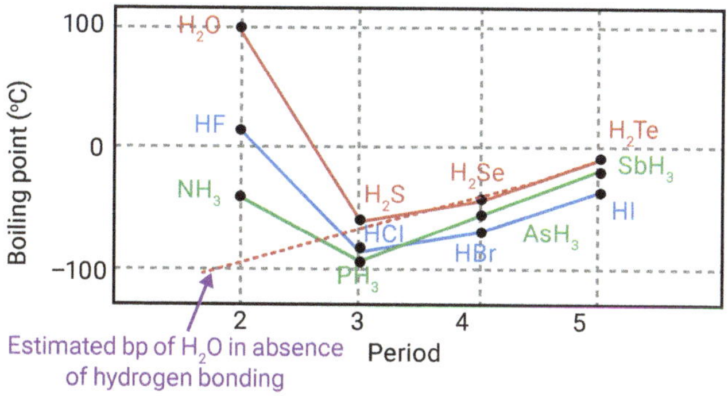

AsH_3—arsine, H_2Se—hydrogen selenide, H_2Te—hydrogen telluride, HBr—hydrogen bromide, HCl—hydrogen chloride, HF—hydrogen fluoride, HI—hydrogen iodide, NH_3—ammonia, PH_3—phosphine, SbH_3—stibine

Figure 2-3 Chart showing boiling temperatures of various hydrides including water, and also the expected boiling point of water if it was not affected by the hydrogen bond

it will have dissolved many of the minerals with which it comes into to contact. This will, in turn, affect the amounts of chemicals we use to treat it, the amount of hardness it exhibits, the degree of corrosivity it will have on our piping and metal tanks in distribution systems, and how our customers view its relative potability.

Water—unlike other hydrogen-bearing molecules—has a greater range of temperatures at which it will remain as a liquid before it either boils off or solidifies into ice.

For example, in Figure 2-3, the red line depicts that various hydrides boil at temperatures less than 0°C, but water has to be heated to 100°C before it will boil—the laws of chemistry would not have predicted that. This is one of those exceptions to the rule. The other molecules on the red line have elements in them from the same column on the periodic table, but they don't act like water does at similar temperatures. This is the "magic" of the water molecule with its hydrogen bond. Without that H bond, chemists believe that water would boil at much lower temperatures.

Water also has many physical properties associated with it that are important and that affect treatment. For example, operators know that warmer temperatures in water will drive reactions to occur more quickly, so it might affect coagulation. Total ionic strength will play an important role in corrosion control programs.

In general, these physical properties are easily measured by operators with standard equipment that is found in even small water system labs. Important information like temperature and conductivity should be recorded in a journal each day so that a record of the trends is kept. Physical data like these can often augment or help explain some of the more sophisticated data that are generated. In the end, operators will choose a treatment strategy based largely on the chemistry of their source water.

SOURCE WATER DICTATES TREATMENT

The treatment challenges operators face often start with the complexity and availability of their source water. Some operators struggle to find enough of it to satisfy the needs of their communities. Others struggle with plentiful source water that Mother Nature has seen fit for use as a reservoir of undesirable minerals that must be removed. Some sources, although once acceptable to a community as a water supply, have become so polluted by human activity that the water supplier deems it necessary to seek alternative sources. When forced to do so, operators must take care to study the probable consequences of switching source waters. The lead (Pb) and *Legionella* crisis in Flint, Mich., was due in large part to the attempt to switch from one source to another without adequate preparation. Another infamous source water disaster took place in Walkerton, Ont., where a lack of operator training and an unprotected source water combined to bring illness and death to that community.

Source water assessment is crucial when choosing treatment options that are designed into a WTP. The reliability of the processes that are ultimately chosen will be based on the historical water quality of the source when available. If the source is newly developed, great care must be taken to characterize it before the design concepts are realized. The resulting design must provide the WTP with the capability to provide consistent water quality that meets all applicable regulations and standards or goals.

Source water management techniques help operators because water management tends to minimize problems that may occur later in the treatment plant. It is often easier and more economical to protect the supply source from contamination than it is to deal with the contamination after it enters the plant. Source water protection, which is discussed later in this chapter, is the first barrier of a multiple-barrier approach to water quality. Operators should refer to Chapter 7 ("Source Water Protection") of AWWA Manual M50, *Water Resource Planning*, for making plans to protect their source.

SOURCES OF WATER

Groundwater

Water that has percolated into the soil or seeped into the ground from streams and other surface sources is called *groundwater*. Water will continue to seep downward into soil until it reaches an impervious rock or clay formation. It then fills the voids of the soil or sand above the impervious cap and is referred to as an *aquifer*. If more water percolates through the soil, excess will accumulate in a zone of saturation, the top of which is called the *water table* (Figure 2-4). As long as water continues to fall into this area faster than it is taken away, the aquifer will yield a source of supply. Aquifers can be

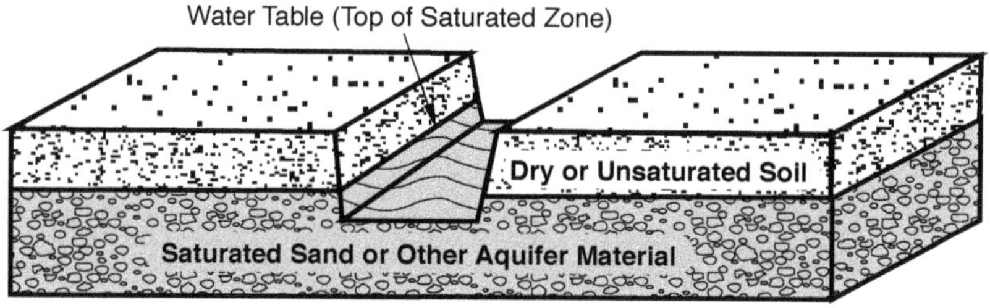

Figure 2-4 Schematic of a water table showing saturated zone

quite large, holding billions of gallons of underground water. Nevertheless, large cities and populations are capable of using this source so quickly that the water table can drop. In arid regions of the southwestern United States, this is a serious problem as the population migrates to these areas and makes more demands for water. Without adequate rainfall to replenish the aquifer, supply dwindles, and conservation measures must be implemented.

Excessive withdrawal may affect the water quality of the aquifer. Examples are chronic seawater infiltration in the New York and New Jersey areas, as well as coastal Florida and California. Excessive amounts of iron (Fe) and manganese can also be drawn into the influence of a well if it is operated at higher than normal rates.

Groundwater occurs in two different kinds of aquifers: *unconfined* and *confined*. An unconfined aquifer, sometimes called a *water-table aquifer*, has an upper surface that is free to rise and fall with incoming and outgoing water (Figure 2-4). Wells constructed to draw from these sources can vary in their ability to yield a specific flow. A confined aquifer has an upper impermeable layer or a layer with low permeability. These aquifers are recharged at an elevation higher than the main portion of the aquifer. They are called *artesian aquifers*, and the water in them is usually under pressure. When these pressurized water tables intersect with the ground surface, they are called *springs*.

The nature and physical properties of the aquifer are important to the operator. The soil that holds the water will be porous to some degree. The more porous the soil, the more it will hold per unit volume. Also, the size of the soil particles will greatly influence the water's ability to move through the aquifer. Large-grained, sandy aquifers will allow water to move more freely than smaller-sized sands, and pumping costs will be less. Some aquifers are made of limestone or fractured rock, and channels develop in them through which water can flow.

Although it is tempting to think of groundwater as free from microbial contaminants, it can become contaminated in many ways or come under the influence of contaminated surface waters. In 1989, the Surface Water Treatment Rule (SWTR) mandated that groundwater under the influence of

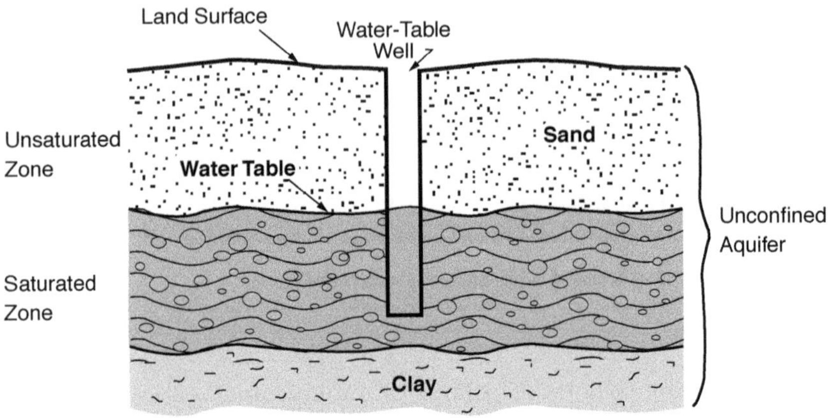

Figure 2-5 Schematic of an unconfined aquifer

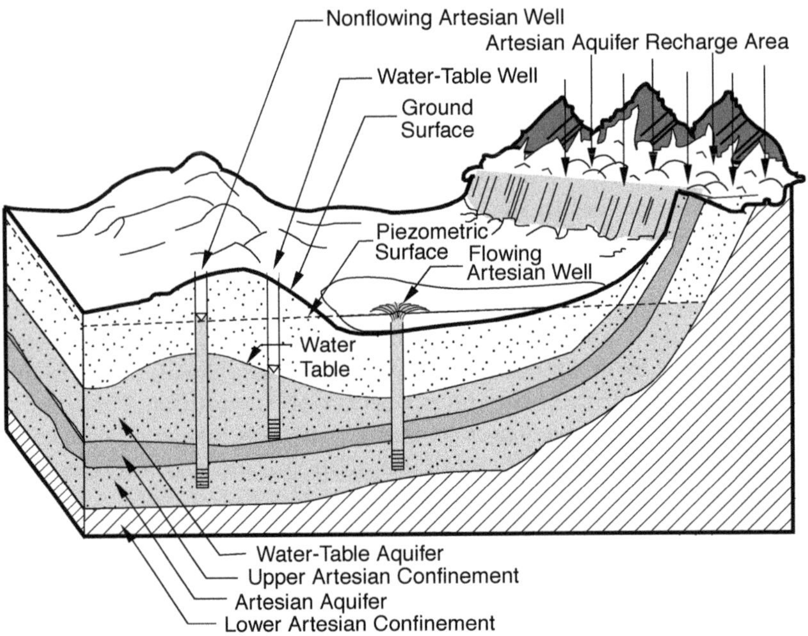

Figure 2-6 Schematic of an artesian aquifer

surface water be treated similarly to surface water and that operators provide treatment techniques for microbiological contamination. In May 2000, the US Environmental Protection Agency (USEPA) proposed the Ground Water Rule, which targets the nation's groundwater supplies. It passed into legislation in October 2006. The USEPA is determining, among other things, the amount of risk because of fecal contamination of these sources. Systems that are shown to be at risk for fecal contamination will be required to provide 4-log inactivation of viruses (99.99% removal). The USEPA is conducting surveys of the more than 147,000 groundwater systems in the United States and is looking at eight critical components to protect public health. Corrective actions are required for any system with a significant deficiency or source water fecal contamination. The system must implement one or more of the following correction action options:

- Correct all significant deficiencies.
- Eliminate the source of contamination.
- Provide an alternative source of water.
- Provide treatment that reliably achieves 99.99% (4-log) inactivation or removal of viruses.

The 1986 Amendments to the Safe Drinking Water Act (SDWA) called on each state to develop a wellhead protection program (WPP). This legislation established a national program to encourage states to develop systematic and comprehensive programs to protect public water supply wells fed by groundwater and well fields within their jurisdiction from contamination. (Certain WPP components are required to be implemented by states under the Source Water Assessment Program, which was a part of the 1996 SDWA Amendments.) States encourage municipalities to participate and to exercise individual discretion in developing methods for protecting groundwater used for drinking water. The Source Water Protection Program (SWPP) incorporates surface water systems and is very similar to the WPP.

Watershed Sanitary Surveys

Sanitary surveys were discussed in Chapter 1. However, some states require separate surveys that are focused on watershed issues. These are often incorporated into broader sanitary surveys but not always. Watershed surveys assess the adequacy of supply, examine the potential for contamination, and may suggest protection strategies. Determining the capacity to sustain a viable, safe water supply is a major outcome of these surveys.

Surface Water

Water found in lakes and streams is referred to as *surface water*. The availability and quality of this source of supply vary greatly by location. All surface supplies are assumed to be microbiologically or biologically contaminated to some degree and require some form of treatment before human consumption. The source for most surface supplies comes from precipitation—either rain or snow. Because the atmosphere contains a considerable amount of suspended impurities from volcanic activity and wind-blown soil, rain and snow become contaminated as they fall to earth. The amounts and types of impurities will vary by region.

When precipitation collects in lakes and streams, it becomes further contaminated as it travels over the terrain. It dissolves hardness-bearing minerals and can be subject to runoff conditions that may bring biological constituents to the water. As it passes over areas with organic content, the water collects natural organic matter (NOM), which is a precursor to disinfection byproducts (DBPs).

Surface water can be soft or hard and may require different treatment for this characteristic. All surface water supplies require filtration and disinfection unless they can show certain avoidance criteria for filtration, as outlined in the SWTR and Interim Enhanced Surface Water Treatment Rule (see Chapter 1). All surface supplies must be disinfected.

Water Impoundment Management

Some utilities construct impoundment reservoirs or use water from naturally occurring reservoirs that are fed from surface streams or that catch precipitation. Constructed reservoirs serve several useful purposes. They can store amounts of water in excess of daily demand and therefore act as a hedge against drought conditions. They can serve as pretreatment reservoirs when aeration systems or chemical treatment are added, or they may be used simply as large settling ponds before treatment. Reservoirs also allow the operator to draw water from nearby streams for storage. In this way, water can be pumped from the stream on days when water quality is excellent. When quality is poor, such as after a rainstorm or when there is an upstream spill, pumping can be discontinued if the reservoir is large enough to store sufficient water for treatment. When conditions return to normal, pumping from the stream can resume.

Reservoirs can also be a source of water quality problems because they are exposed to the elements. Sunlight penetration into the depths of the reservoir can promote algal growth, which causes taste-and-odor problems. Runoff from the ground can bring with it chemical and biological contaminants. Extreme wet and dry conditions affect water quality. For example, heavy precipitation can resuspend bottom sediments and increase turbidity. This can load the treatment plant with huge amounts of solids, making treatment expensive and management of residuals difficult. Drought conditions can cause stagnation and minimize the otherwise healthy dilution of contaminants because of low water levels.

The depth of a lake or reservoir is subject to two conditions that can affect source water quality. The location of the intake with respect to these depths is important. The first condition, known as the trophic cycle, involves three stages at which nutrients and biological activity influence water quality. The *oligotrophic* stage is associated with low nutrients and limited algal production and biological activity. The *mesotrophic* stage involves a moderate amount of these constituents, and the *eutrophic* stage is rich in nutrients and biological activity. This eutrophic stage is associated with low oxygen levels, high turbidity and color, and NOM.

Because oxygen is depleted, usually by algae, an anoxic condition can develop. This creates a reducing environment, where insoluble iron and manganese bottom deposits can become soluble ions. If this water is brought

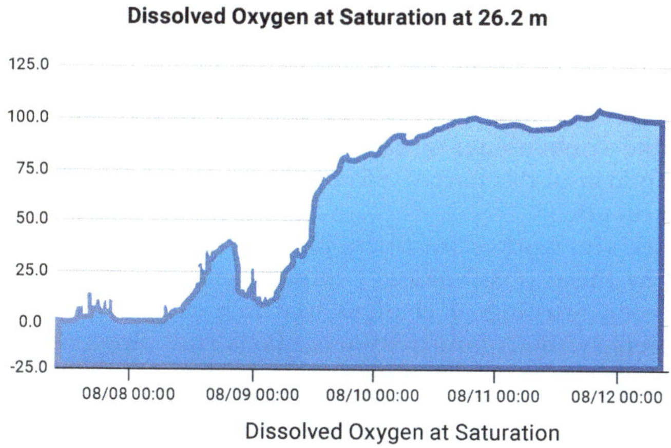

Source: Nick Pizzi

Figure 2-7 Dead zone water graphic showing oxygen depletion

into the plant, water quality problems can occur that can require treatment. In the following graphic (Figure 2-7) from a Lake Erie WTP, we see the dead zone water with low oxygen levels occurring for two to three days. Neighboring utilities showed similar results. During this time, the source water temperature dropped from 26 to 17°C in less than an hour. The source pH dropped from 8.3 to 7.5, and manganese levels increased along with chlorine (Cl) demand. Manganese at even low levels takes a lot of oxidant and time to turn from soluble to insoluble at these low pH values.

The second condition that can affect source water quality is *thermal stratification*. Layers of the lake or reservoir, called *limnion*, develop because of temperature differences. Oxygen depletion can take place within these layers and cause problems similar to those described previously. In colder climates, water at the top of a lake or reservoir can cool to around 39.4°F (4°C), a temperature at which water reaches its greatest density. This cooler, denser water sinks and forces the warmer water at the bottom to the top. This mixing can be so violent that the bottom of the lake or reservoir is riled, causing poor-quality, high-turbidity water to be brought into the plant. Operators refer to this phenomenon as *lake turnover*. It can occur each fall and spring. Some operators have witnessed this phenomenon in uncovered sedimentation basins.

Riverbank filtration is considered groundwater under the direct influence of surface water. Treatment plants with this source of supply receive water of changing quality, similar to the river, but often the quality is more stable. The filtration of the soil either in the bottom of or adjacent to the river provides beneficial pretreatment. In some situations, the degree of pretreatment has been assessed, and the regulatory agency has granted some treatment credit to this process. This is site specific and at the discretion of the regulatory agency.

Direct river intakes often challenge water supply sources. Although there are advantages to drawing water directly from a river, such as contamination incidents that may flow downstream quickly beyond the intake, water quality can change rapidly compared with supplies from impoundments. Many possible contaminant situations may affect water quality: chemical spills from industrial discharges, river shipping accidents, floods, droughts, and surface runoff.

Operators who work at treatment plants with this source of supply need to be alert to these possibilities. Online, continuous monitoring is often used to provide warning of changes that may require adjustments to the water treatment system. Many treatment plants with direct river intakes have procedures in place to shut down production if certain conditions develop. Protecting the public water supply is of primary importance.

PARTICLES AND NOM

The conventional surface WTP operator spends a lot of time and energy removing two source water constituents: particles and NOM. There are two major rules in the SDWA that demand it. The SWTR requires removal of particles as measured by turbidity, and the Disinfectants and Disinfection Byproducts Rule (DBPR) requires the removal of total organic carbon (TOC) as a precursor to DBPs.

Particles

Particles found naturally in source water are important because they can affect treatment processes and the finished product. Some particles can adversely affect the aesthetic qualities of water, while still others can affect human health. Particles can also inhibit the disinfection process. Particles can foul membranes, granular activated carbon adsorption sites, and ion exchange media. Of course, some of the treatment processes that are used to remove particles create additional particles on their own. Coagulation, flocculation, precipitative softening, and oxidation of metals such as iron and manganese are all processes that deliberately create floc particles that must be removed.

Source water particles are found in all shapes and sizes, living and inert, organic and inorganic, with material-dependent densities and often a surface area that carries a net force that can easily repel other particles of similar net force. When these particles are suspended in water with other electrostatically charged particles, they will remain in that stable state until they become destabilized by processes like coagulation and flocculation. Any particle not destabilized won't readily stick to a grain of filter sand or at the surface of a hollow-fiber membrane.

It's common to think of particles as tiny spherical objects floating in the water, when in reality many different shapes are seen. Shape can affect processes even where pretreatment is good, such as when rod-shaped or

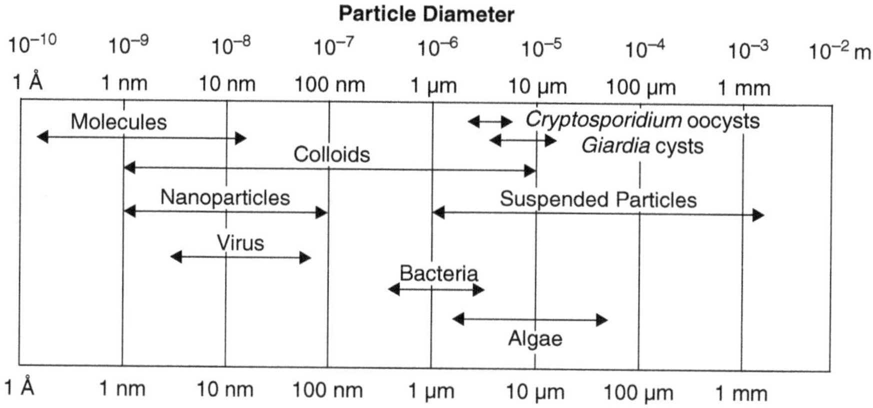

Source: AWWA 2011

Figure 2-8 Size range for particulate matter in water

filamentous algae clog filters, or when agglomerated floc becomes too wispy and refuses to settle quickly.

Size is thought to be the most defining characteristic of a particle. From Figure 2-8, we can see that particles range from floc particles at 1 or 2 mm down to viruses and colloids that are five or six times as small. We can deduce from the chart that many of the small particles—especially the ones we don't create when we add coagulants—are also considered to be colloids and likely to have the negative charges that must be destabilized.

Operators take advantage of two important characteristics of particles in their treatment practices. One is the ability of particles to scatter light, which allows for turbidity (and particle count) measurement. The other is the tendency of particles to accumulate electrical charge on their surface, which allows for measurement of the strength of particle stability. The stronger the charge, the more likely the particle will resist settling. Using streaming current detectors, operators can determine the lowering of this resistant charge. Because particles often aggregate (flocculate) well at a net-zero charge, dosage of coagulant can quickly be adjusted.

NOM

NOM in source water is important to operators because it can impart color to the water, react with disinfectants and oxidants (increase chlorine demand), and foul membranes and other treatment processes such as powdered activated carbon adsorption. NOM can be particulate (e.g., algae) or dissolved (e.g., decayed vegetation). The NOM from decay reactions of plant and animal matter in the watershed is composed of humic matter. Runoff events wash this matter from farmlands, bogs, and swamps into the source water, and it mixes in with nonhumic NOM. Nonhumic matter is not removed as well as humic matter by coagulation and flocculation.

TOC is a measure of all of the organic carbon in the water and is composed of the particulate organic carbon (POC) and dissolved organic carbon (DOC). Measurement of DOC is an operational definition because samples are filtered through 0.45-mm-pore-size filters to obtain the fraction of organic carbon. DOC is defined as the carbon that passes that filter. Therefore, for operational purposes, POC is defined as the TOC minus the DOC. In surface waters, the POC may be approximately 1–10% of the TOC. In groundwater, POC is usually zero, so the DOC is equal to the TOC.

As stated in the first chapter, TOC removal is regulated, and the plant operator is required to remove a percentage of it based on source water alkalinity. According to the enhanced coagulation requirements of the Stage 1 DBPR, samples must be taken periodically to determine compliance. If compliance levels cannot be met, there are criteria that can be followed. Removal of TOC is important because it reacts with chlorine or other oxidants. The operational goal is to remove as much of the TOC as possible before chlorine addition.

But as an operational tool, TOC testing is not possible for many small and midsize utilities. Fortunately, there is a surrogate for TOC—ultraviolet absorbance at 254 nm (UV254)—and it is easily measured in the lab with very little effort or expense. This is important because it represents that portion of the highly charged (negative) NOM that can be removed efficiently with positively charged alum or ferric coagulants used in most conventional plants. Because it is easily measured onsite, the operator can quickly adjust dosage of coagulant based on a target UV254 reading.

UV254 is absorbed readily by organic molecules, especially those humic substances that are precursors for total trihalomethanes. The specific UV254 (SUVA) calculation further defines the portion of NOM that has the highest absorbance and therefore can be targeted by operators for treatment. Over time, the operational staff can correlate UV254 with raw and finished water TOC levels. Figure 2-9, taken from *Water Quality and Treatment* (AWWA 2011), shows the data that one utility has accumulated over a five-year time frame that allows them to predict SUVA value and expected TOC removal.

SOURCE WATER PROTECTION PROGRAM

A utility's watershed management program is the first barrier of protection against pathogen intrusion. Such a program can also protect drinking water quality by minimizing DBP precursors and protecting against algal growths that cause taste-and-odor problems. Deterioration of a watershed not only results in more costly treatment requirements at the plant but also can erode public confidence. If a consumer perceives that the watershed providing drinking water is not managed well, the consumer may seek alternative sources.

In an ideal setting, water utilities would own all of the land that comprises the watershed for their source. In fact, land ownership was the principal

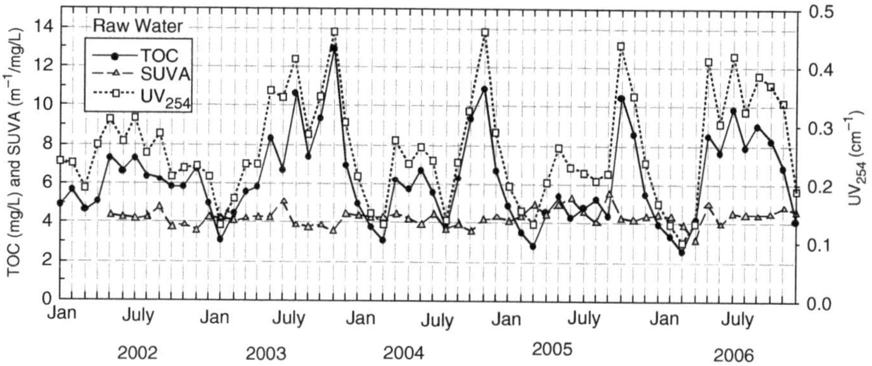

Source: AWWA 2011

Figure 2-9 Chart showing comparison of TOC, UV254, and SUVA values in source water

method of protecting source water in early protection programs. Currently, some utilities can still claim virtual ownership of their source's watershed; New York, Boston, Portland, Seattle, and San Francisco are among those. These utilities have avoided filtration requirements in part because of the high quality of their source water, which is a direct result of extensive land ownership and watershed management programs.

However, there is concern about microbial contaminants that are resistant to chlorination. By understanding the watershed and its sources of contamination, both active and potential, drinking water utilities can become advocates for improving their watershed and source water. Using tools such as those provided under the Clean Water Act to achieve water quality goals, operators can develop a watershed protection program. In effect, Storm Water Pollution Prevention (SWPP) plans are simply an old idea that has received new attention to help guard against "new" threats. In general, a good watershed management program contains the following components:

- Identification and characterization of the source
- Characterization of potential effects on the source
- Determination of the intake's vulnerability to contaminants
- Established protection goals
- Development of protection strategies
- Program implementation
- Program monitoring and evaluation

Many utilities have initiated SWPPs. An excellent reference book for source water issues is AWWA M50. In Chapter 7 of M50, which addresses source water protection, we find the following quote:

> Throughout most of human history, protecting drinking water supply sources was the only viable option for obtaining safe drinking water. In modern times, even with advanced technology treatment methods that many water systems rely on for producing safe drinking water, source

water protection programs are often considered the preferred approach rather than using water treatment technologies to remove contaminants such as anthropogenic compounds, like synthetic organics and chemicals, and natural contaminants like arsenic and pathogens.

In other words, it's usually better to stop source water contamination than to rely on the WTP to suddenly have to deal with it.

Modern-day drinking water disasters that took place in Flint, Mich., and Walkerton, Ont., were preventable source water issues.

SOURCE WATER QUALITY CONSIDERATIONS

Many possible water quality issues confront water treatment facilities. Recognizing these challenges and being prepared to effectively adjust to them is the primary task for WTP operations personnel. It is the reason for extensive education and training. When the water quality changes and the treatment plant systems must be adjusted, experience and training are called on to meet the threat.

A treatment challenge is presented when the source water quality *changes*. A steady predictable water quality, even if it is not the greatest, can often receive effective treatment. The treatment system can be refined over time, and excellent drinking water quality can be produced from even highly impaired waters.

Both surface and groundwater can change and potentially create a situation in which water quality is at risk. Some of the water quality issues that may result from changes in source water quality are

- bacteriological—e.g., fecal contamination;
- biological—e.g., zebra mussels, algae, and algal toxins;
- chemical—e.g., hardness, per- and polyfluoroalkyl substances, radioactive substances, and spills;
- physical—e.g., temperature and pH;
- agricultural—e.g., pesticides and fertilizer;
- groundwater conditions—e.g., wellhead contamination; and
- lake stratification—e.g., dead zones, taste and odor, and iron or manganese.

Direct river intakes can be very challenging water supply sources. Conditions can change quickly, so operators must be alert to these conditions. Treatment for known changes should be documented and plant adjustments rehearsed (tabletop exercises) to ensure that acceptable drinking water quality is maintained. This is especially important at facilities where water quality seldom changes, where operators can become complacent with the everyday routine and be caught off guard.

Lakes or other impoundments often stabilize water quality for lengthy periods. However, changes can occur, such as lake stratification and turnover. Another common cause of water quality changes in lakes is the growth of algae and other aquatic organisms. Some of these organisms can create

Source: Aqua Ohio–Struthers Division

Figure 2-10 The dam at Evans Lake—source water for the Struthers, Ohio, WTP

disagreeable taste and odors. Others may contribute chemical byproducts that can be toxic. Treatment of these conditions is best performed in the lake and controlled, if possible, to avoid these problems. Control methods for nutrients such as nitrogen and phosphorus (P) compounds have been used to reduce algae and aquatic weed growth. In most instances, however, chemical control of algae is necessary. Copper sulfate is often the chemical used at a generic application rate of 1.1 lb/ac-ft. Also, strategies have been used to reduce stratification by mixing or oxygen (air) application. These practices are often successful in reducing potential water quality problems but not eliminating them completely.

As mentioned earlier, in the case of surface water "dimictic" lakes such as Lake Erie, dead zones and turnover are a common problem. These lakes have two periods of mixing and turnover (spring and fall); they are stratified in the summer, and then they freeze in winter and become inversely stratified. In later summer and early autumn, dead zone water, which is very cold and devoid of oxygen, tends to remain intact and not mix with warmer oxygenated water. WTPs that are unfortunate enough to take in this water can bring in chemically reduced iron and manganese. Manganese in this form and at low pH values is very difficult and time-consuming to remove.

Groundwater sources can be contaminated by surface runoff entering an unprotected wellhead. This sort of incident can be serious because the well may not be monitored for this unforeseen occurrence. Disinfection strategies that include continuous monitoring and pacing chemical addition can be an effective defense when demand increases expectantly.

Source: Aqua Ohio, Massillon

Figure 2-11 Well #17 at Massillon, Ohio, water system

Understanding the characteristics of the water supply source and preparing treatment practices to combat quality changes are the best ways to maintain high-quality water production. Treatment plant systems must be adjusted to meet changing water source conditions. WTP operators should have a proven plan in place that does not depend on automatic control systems such as supervisory control and data acquisition to deal effectively with water source quality changes. Periodic sessions of "manual operation" are a useful tool to help keep operators sharp.

BIBLIOGRAPHY

AWWA. 2017. M50, *Water Resources Planning*. Denver: AWWA.

AWWA. 2011 (6th ed.). *Water Quality and Treatment* (J.K. Edzwald, editor). New York: McGraw-Hill.

AWWA. 2010a. M57, *Algae: Source to Treatment*. Denver: AWWA.

AWWA. 2010b (4th ed.). *Principles and Practices of Water Supply Operations—Source Water*. Denver: AWWA.

Pizzi, N.G. 2024. "Lessons From the Past Can Help Operators Protect Public Health." *Opflow*, 50(4):10-14.

Chapter 3

Well Design and Operation

Exploration and site selection are important issues when developing well-water sources because each well system is unique. Optimal production and use of a well source are possible only by planning in detail, properly testing and evaluating the data gathered, properly selecting pumping equipment, and properly operating the well field.

A well should be located so as to produce the maximum yield possible while still being protected as much as possible from contamination. State and federal geographical and specialized geological and groundwater databases, which include information about water quality and availability, should be surveyed so that pertinent data can be examined for site location. Generally, hydrogeologists should be part of the planning team from this point because of their specific knowledge of groundwater systems. Other information sources include landowners in the surrounding area and local commercial well-drilling contractors. Commonsense examination of the topography can lead to discovery of a well-water source. It is likely to be present under valleys rather than under hills, and it may exist where greenery is lusher and where streams or springs abound. Exploration methods such as geophysical testing, computer modeling, and test drilling might be necessary to confirm suspicions that water is present.

The US Environmental Protection Agency (USEPA) has implemented the Ground Water Rule, which requires proof of disinfection (for many groundwater systems) daily for the water drawn from the well before distribution to customers. Well-water system operators must comply with the requirements of *concentration* times *time* ($C \times T$) that surface water systems have used for years. Some groundwater supplies have been granted a waiver from the filtration and disinfection requirements. These systems must monitor for certain contaminants to continue to qualify for the waiver.

TYPES AND OPERATION OF WELLS

In considering a groundwater-supply system, there are two major components: the geologic system, including the aquifer that stores and transmits water, and the constructed well that serves as the access point to the aquifer while, ideally, keeping undesirable water and earth materials out of the water supply system. A well functions in a way similar to the way a dam or reservoir captures surface water. An aquifer stores water, while a well provides a way to access the reserve of water. The well pump is not unlike the dam gate. It controls the flow of water from the reservoir. All wells work similarly in theory, but special considerations and differences are found from one well type to another. The following is a list of well types.

Dug Well

The oldest type of well, going back tens of thousands of years, a dug well is capable of providing large amounts of water from shallow groundwater sources. They are literally dug by hand with pick and shovel; larger wells are dug with a backhoe or crane and clamshell. After excavation, the hole is lined with bricks, stones, cement, and other structural materials if the soil is unstable. Because they are large in diameter, dug wells are more difficult to protect from surface contamination. In modern times, these are rare in large-scale water supply but may be used in groundwater recharge.

Bored Well

With suitable soils, a well can be developed quickly by boring with an auger. These wells are limited to a 3-ft (1-m) diameter and may penetrate 25–60 ft (8–19 m). A variation is the bucket-auger method, which can drill 3–4 ft (1–1.3 m, sometimes larger) in diameter, and somewhat deeper. Bored wells are not common in public water supplies.

Driven Well

Driven wells consist of a pointed well screen, which is called a drive point, and lengths of pipe, which are attached to the point. Driven wells usually do not exceed 4 in. (10 cm) in diameter, nor do they probe deeper than 40 ft (13 m). For these reasons, they are not often used for public water supplies. They may have use as temporary or even permanent monitoring wells.

Jetted Well

Where soil permits, a special cutting tool that uses high-pressure water pumped through a cutting knife is used to "jet" a well hole. The casing is attached to the jet and is laid in place as the unit cuts into the ground. This system is not often used for public water supplies, although historically common in glaciated New England in the United States.

Drilled Well

Drilling of a well is the most common type of well development method for the public water industry. The method is versatile, can accommodate extreme depths, and can produce diameters of 4 ft (1.3 m) or greater, but more typically 1–3 ft (up to 1 m). Drilling methods include

- cable tool method;
- rotary hydraulic method;
- reverse-circulation rotary method;
- California method;
- rotary-air method; and
- down-the-hole hammer method.

Each of these has variations—for example, the increasingly common dual rotary method that rotates in a casing (for construction or permanent) while drilling. Other specialized methods such as rotary sonic or coring are used for test drilling.

With some of these methods (bucket auger, cable tool, dual rotary, and air rotary with casing advance), drilling may advance a construction casing to the desired depth (usually defined by test drilling); then, a well is built inside of this borehole, and the construction casing is removed.

In each case, drilling is completed by development that removes fine particles and drilling damage to improve the hydrologic contact between the well and aquifer, reflected in improved specific capacity.

Radial Well

Radial wells are special service wells that are commonly used near the shore of a lake or river to obtain large amounts of relatively good-quality water from shallow, often thinner aquifers. A large-diameter caisson that may be able to house multiple pumps is excavated and sunk as in well digging to the target depth; then, laterals (essentially horizontal screened wells) are pushed out into the formation at desired depths, directions, and lengths (defined by test drilling). Water quality can improve as the water travels through the sand or gravel bed in the stream and into the well. There is an interest in these wells because utilities may receive some removal credits for contaminants through the system. In some jurisdictions, engineered filtration and disinfection may be avoided altogether. With this "riverbank filtration" credit, a utility may be able to avoid other, more expensive treatment technologies. Even without a regulatory benefit, they greatly simplify treatment by providing a filtered source water.

PARTS OF A WELL

Figure 3-1 shows the parts of a typical, relatively small, drilled well. The well is equipped with a sanitary seal that prevents contamination from entering the well casing. The seal has openings into the well for discharge piping, air vents, and pump controls. Air vents let air in as the well level drops during operation. The well casing is a liner placed into the borehole to keep the walls from caving in. Grouting is cement (sometimes specialized bentonite)

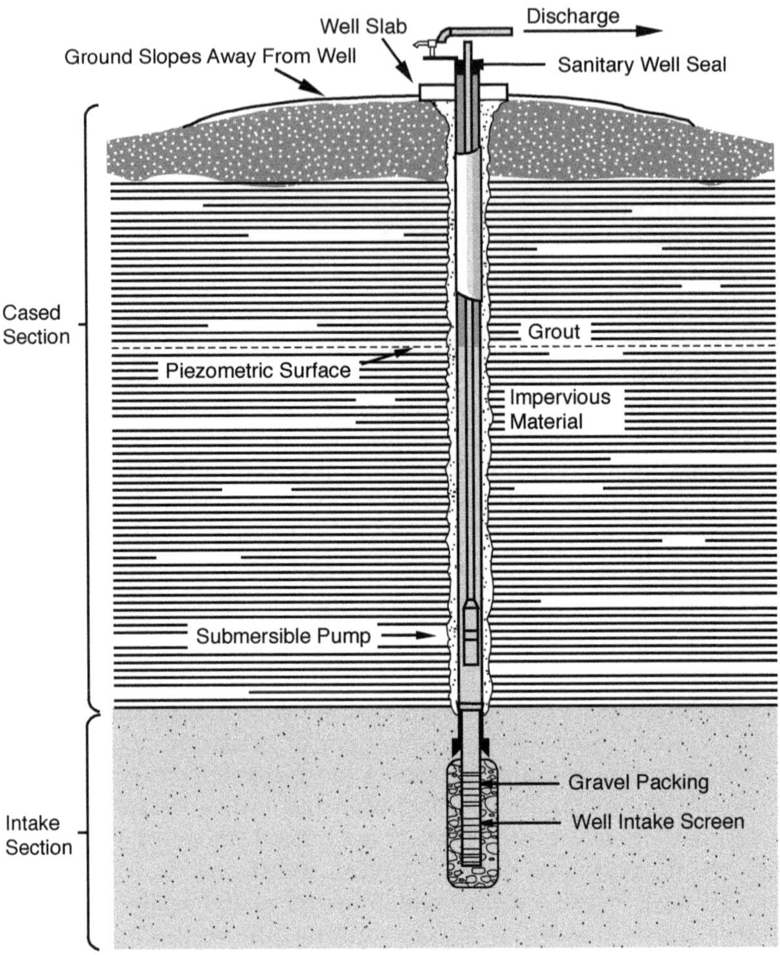

Figure 3-1 Parts of an underground well

that is pumped into the space between the drilled hole and the casing to protect the casing and prevent water from traveling along the casing from the surface or between aquifers. The well may have a screen if the source is in sandy or gravelly aquifer formation or unstable aquifer rock, or where screen and filter pack are required of all wells' soil.

OPERATIONS AND MAINTENANCE

Operators of well systems must be familiar with special terminology. These terms are listed in this section.

Static Level

The static water level (sometimes "resting dynamic level") is the water surface in the well when no water is being pumped or taken from the well and adjacent aquifer. It is usually measured as the distance from the ground surface to the water surface. This measurement is used to monitor changes in the water table and should be taken and recorded frequently. The static water level in a particular well may be subject to variation because of nearby pumping wells, seasonal effects, and long-term trends.

Pumping Level

The water level in a well usually drops when the pump is operating and water is being removed. When the water stabilizes at this lower level, it is measured as the pumping water level or dynamic water level. It is important that the pump intake be located below this level.

Drawdown

Drawdown is defined as the difference between the static and pumping water level, as shown in Figure 3-2.

Cone of Depression

In an unconfined aquifer, the flow of water in the aquifer to the well comes from all directions surrounding the well. The water surface in the aquifer takes on the shape of an inverted cone, known as the cone of depression (Figure 3-2).

Radius of Influence

The zone affected by drawdown extends outward from the well at a distance that depends on several factors, including aquifer permeability, aquifer

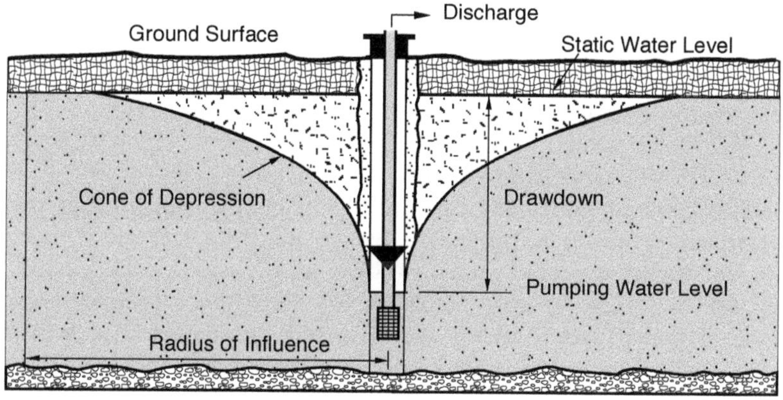

Figure 3-2 Drawdown in a well

thickness, sources of or barriers to recharge, whether the aquifer is confined (under some pressure or restriction) or unconfined (water table), and the porosity of the soil. The steepness of the cone corresponds to a radius of influence that extends horizontally outward from the well. Water will move easily through an aquifer composed of very sandy soil, and the cone will be almost flat. A more impervious soil will create a flow that deepens the cone and extends it out further. This is important to operators because wells can be situated near each other. If the radius or zone of influence of two wells overlaps, the wells will affect each other's yield.

Well Yield

The yield of a well is the rate of water withdrawal that a well can supply over a long period of time. In conventional US usage, yield is measured in gallons per minute for small wells but may be measured in cubic feet per second for larger wells. Globally, yield is measured in liters per second, minute, or hour or cubic meters per hour or day. When the pumping rate from the aquifer exceeds the water recharge over time, the drawdown will gradually extend to greater depths, and the safe well yield will be reduced. This can allow the pump to cavitate or suck air, which can potentially damage the unit. Operators then either lower the pump itself (obviously up to limits) or operate the pump for shorter periods of time, allowing the well to rest and recover. Best of all would be to operate the well within sustainable constraints that can be designed by hydrogeologic analysis.

Recovery Time

Recovery time is the time that a well needs to rest to come back to safe yield.

Specific Capacity

Operation of a well system requires testing for specific capacity, which is a measure of the well yield per unit of drawdown. The formula for specific capacity is

$$\text{Specific Capacity} = \text{Well Yield/Drawdown} \qquad \text{(Eq 3-1)}$$

This example illustrates use of the formula. If well yield is 200 gpm (0.44 ft^3/s) and the drawdown is 25 ft (8 m), the specific capacity is 200/25 = 8 gpm/ft (1 ft^3/min/ft) of drawdown.

Specific capacity will vary depending on changes in aquifer transmissivity (Hydraulic Conductivity × Thickness—just like a filter) if the aquifer water levels decline over an area or region and if the screens or intakes of the well are clogging. Specific capacity has two major components, often known as well (or turbulent) loss and aquifer (or linear) loss. Well testing can distinguish the amount of loss in these components to help in planning responses.

Surging

Fine particles that accumulate on the well face and reduce efficiency must be removed. This is done using a plunger, or surge block, that is moved up and down the well casing; the process is called *surging*. This maintenance procedure is performed periodically to ensure well efficiency. The process allows for the fines to fall to the bottom of the well, where they are removed with a bailer. Some well operators use air for surging. High-rate pumping coincident with surging may also be used to improve efficiency during the drilling process. High-rate pumping alone may also be used to clean out the casing in a rock aquifer.

WELL OPERATIONS

Well level, pressure, and flow are some of the minimal operational data that should be kept. These are captured in trend charts on the supervisory control and data acquisition system at the Aqua Ohio–Stark Regional Water Treatment Plant (WTP), as seen in Table 3-1. The Stark WTP is a precipitative softening plant that uses lime to treat a design flow of 15 mgd. The plant operates 24 hours per day and treats about 10 mgd. The source for the plant is well water drawn from 11 wells located in three well fields north of the plant.

Operators record flow data daily on a spreadsheet for each well. It is important for both the operators and the maintenance employees to routinely look back through the trends to ensure that no issues have occurred with the level sensors, pressure transmitters, or flowmeters. On an annual basis, these are tested manually, and specific capacities are tracked to determine when the wells should be scheduled for cleaning or maintenance. In Table 3-1, you will see the flow that was recorded for July 2024. The specific capacity for a well can be determined by the flow by the amount of drawdown in a well (flow/drawdown-static). This information should be used as one of the components in determining well maintenance.

Disinfection of Wells

Disinfection of wells is part of the startup and maintenance procedures that operators must follow. These techniques are discussed in Chapter 8 and Appendix A.

Operators should refer to AWWA Standard C654, *Disinfection of Wells*, for further reference.

Because the Ground Water Rule requires 4-log inactivation of viruses when using chemicals for disinfection, some groundwater systems will need to alter their disinfection processes. Groundwater system operators must understand the effect of these changes on distribution system water quality.

Table 3-2, which was published in the USEPA simultaneous compliance guidance, can be used as guidance to determine possible simultaneous compliance issues that may arise.

Table 3-1 July 2024 well data

Wells	Static Level	Drawdown	Specific Capacity	Min Flow *mgd*	Max Flow *mgd*	Discharge Pressure	Total Flow *mgd*
Well 1	33	67	50	0	1,700	12	75,575,000
Well 9	59	81	45	0	1,000	8	32,979,000
Well 10	61	82	114	0	2,400	6	63,263,000
Well 11	56	78	55	0	1,200	5	19,951,000
Well 12	30	40	120	0	1,200	12	22,595,000
Well 14	38	44	383	800	2,300	18	34,436,000
Well 15	37	41	450	800	1,800	20	29,870,000
Well 16	42	58	113	0	1,800	20	52,211,000

Source: Sean Smith, Plant Manager, Aqua Ohio Stark Massillon WTP

REASONS FOR WELL ABANDONMENT

When a well is no longer of any use, it must be abandoned properly. Wells need to be abandoned properly to

- eliminate a physical hazard;
- prevent groundwater contamination;
- conserve the aquifer; and
- prevent mixing of desirable and undesirable water between aquifers.

In arid regions, well abandonment for aquifer preservation is not uncommon. Well sources that draw from aquifers bordering seawater are sometimes abandoned because continued use of the well would result in saltwater intrusion. In some types of soil, such as that found in Florida, wells have been abandoned because their use produces sinkholes.

Table 3-2 Possible simultaneous compliance issues

Rule or Regulation	Simultaneous Compliance Issue
Stage 1/Stage 2 Disinfection Byproduct Rules	Systems initiating chemical disinfection will need to begin distribution system monitoring.
Stage 1/Stage 2 Disinfection Byproduct Rules	Initiating or increasing chemical disinfection may result in disinfection byproduct formation.
Stage 1/Stage 2 Disinfection Byproduct Rules	Systems using chemical disinfection must meet Stage 1/Stage 2 Disinfection Byproduct Rules' maximum contaminant levels in the distribution system.
Stage 1/Stage 2 Disinfection Byproduct Rules	Systems using chemical disinfection must monitor distribution residual levels and meet maximum residual disinfectant levels.
Lead and Copper Rules	Initiating chemical disinfection may result in changes in water quality parameters, creating the potential for increased corrosivity in the distribution system and increased lead and copper contaminant levels at the tap.
Arsenic and Radionuclides Rules	Initiating chemical disinfection may require changes in corrosion control treatment for water quality parameters, which can affect the effectiveness of some treatments for arsenic and radionuclides contaminants.
Operator Certification	Initiating chemical disinfection may necessitate new operator certification requirements for the system.

Source: AWWA 2010

WELL MAINTENANCE

Periodic maintenance activities should be followed by each water system using wells as a source. The remoteness of most well fields from the plant can produce the "out of sight, out of mind" mentality, and complacency can set in. The staff of the WTP should develop a maintenance program for the wells that requires thorough inspections at least once a year. The oil should be changed in the well motors, and the packing seal should be adjusted on the pump shaft. Electrical motor systems should be monitored and can be used to sense a decline in performance of otherwise out-of-sight submersible pumps. This maintenance is typically contracted out by small systems who will engage a specialist. Larger systems may have a team of operators whose job is to keep the wells operating smoothly.

Records of maintenance activities are important. Once a year (during periods of low consumption), maintenance staff should measure the static and drawdown levels for each well. The operations staff should expect to work with the maintenance activities when needed, especially when turning wells on and off. This way, they will be able to prepare for treatment challenges. Communication is required between the operator and maintenance. Operators should expect to see changes in the source water quality because of the wells being turned on and off. Static and drawdown tests provide data that are important for deciding which wells will be cleaned each year. Some systems with parallel mains coming from the well field will periodically close off one or more of the mains to scour the other mains using the higher velocity. Operators should be prepared for the challenge of poor-quality well water coming in.

BIBLIOGRAPHY

Ahmed, N., S.W. Taylor, and Z. Sheng, editors. 2014. *Hydraulics of Wells: Design, Construction, Testing, and Maintenance of Water Well Systems.* MOP 127. Reston, Va.: American Society of Civil Engineers.

AWWA. 2021. C654, *Disinfection of Wells.* Denver: AWWA.

AWWA. 2014. M21, *Groundwater Operations.* Denver: AWWA.

AWWA. 2011 (6th ed.). *Water Quality and Treatment* (J.K. Edzwald, editor). New York: McGraw-Hill.

AWWA. 2010 (4th ed.). *Principles and Practices of Water Supply Operations—Water Treatment.* Denver: AWWA.

NGWA (National Ground Water Association). 1998. *Manual of Water Well Construction Processes.* Westerville, Ohio: NGWA.

Smith, S.A., MS. 2024. Personal communication.

Smith, S.A. 1999 (5th ed.). "Source Water Quality Management: Ground Water." In *Water Quality and Treatment* (J.K. Edzwald, editor). New York: McGraw-Hill.

Smith, S.A. and A.E. Comeskey. 2009. *Sustainable Wells: Maintenance, Problem Prevention, and Rehabilitation.* Boca Raton, Fla.: CRC Press.

Smith, S.D. 2024. Personal communication.

Sterrett, R. 2007 (3rd ed.). *Groundwater & Wells.* New Brighton, Minn.: Johnson Screens.

Chapter 4

Pretreatment

Pretreatment usually refers to treatment that occurs after withdrawing water from the source but before it enters the first process (usually rapid mixing) in the water treatment plant (WTP). Pretreatment, or preliminary treatment, is often used by water utilities when the source water contains large debris such as sticks and leaves that might damage pumps or other process equipment. Pretreatment is also used when source water contains large amounts of grit, sand, or turbidity. Also, water containing some dissolved minerals or other impaired water quality conditions such as toxic algae may benefit from some type of preliminary treatment. Water that will be treated with membranes can require pretreatment to prevent fouling. Pretreatment may be required continuously or only intermittently depending on seasonal changes.

VARIATION IN SOURCE WATER QUALITY

As discussed in Chapter 2, the quality of the source water will determine the design of the treatment plant and the methodology used to produce potable water. Water quality characteristics fall into four categories: physical, chemical, biological, and radiological. Some of the characteristics are considered contaminants and are undesirable. Operators base the quality of their source water on the concentrations of impurities found in it, and those concentrations vary by source and by characteristic.

One or more impurities will be present to some degree in all waters found in nature. Contaminants of concern include minerals such as per- and polyfluoroalkyl substances, iron (Fe) and manganese, turbidity, color, pH, hardness, organic material, and total dissolved solids. Biological contaminants such as algae, bacteria, zebra mussels, viruses, and *Cryptosporidium* are also a concern. The temperature and taste-and-odor characteristics of the source water will also affect water quality.

Source water quality differs from region to region and also from surface to groundwater supplies. Dissolved mineral concentrations of water vary widely and can contribute to many of the physical and chemical attributes of the water. Table 4-1 is a comparative list of the amounts of some of the

Table 4-1 Typical ionic concentrations in some surface waters

Ionic Species or Parameter	Savannah River	Columbia River	Mississippi River at St. Paul	Missouri River at Hermann	Colorado River Near Phoenix
pH	7.0	(not reported)	8.6	7.7	8.3
Na^+	12	6.2	10	35	(not reported)
Ca^{++}	4	19	23	55	77
Mg^{++}	1	5.1	51	16	29
Cl^-	9	3.5	21	20	88
HCO_3^{-1}	23	76	150	175	135
Hardness ad $CaCO_3$	14.1	68.5	78.1	200	311.8
Ionic strength	0.93×10^{-3}	1.70×10^{-3}	3.30×10^{-3}	3.89×10^{-3}	1.34×10^{-3}
DOC (mg/L)	3.0	(not reported)	8.0	3.8	8.3

Source: AWWA 2010

ions and other parameters found in selected surface sources. Groundwater sources can also vary greatly in their mineral content.

In general, groundwater sources tend to be more stable with respect to temperature, while surface supplies reflect temperature swings because of the local climate. Surface supplies in the northern United States may range in temperature from just above 0 to 28°C or more. Consumers seem to prefer colder water, and industries that use water for cooling need colder water too.

Tastes and odors may intensify at warmer temperatures. The temperature of water in lakes and reservoirs can vary, with colder water generally found at the bottom. In winter in northern climates, this effect may be less noticeable.

In Chapter 2, the concept of stratification was discussed. Large bodies of water can develop "dead zones," where there is little or no oxygen (O_2). These zones may contain reduced (unoxidized) iron and manganese and may be toxic to fish. If brought into the plant, the reduced forms of iron and manganese would exert an increased oxidation demand that may require special treatment. These dead zones can be natural or man-made. The Great Lakes water systems have seen an increase in the frequency and severity of dead zones, especially in Lake Erie, which is shallow and subject to farm runoff that provides nutrients for algae. Operators should note rapid changes in water temperature, pH, alkalinity, and dissolved oxygen as indicators of dead zones.

Climate change has brought rapid depletion of water supplies in some areas and contamination from flooding in others. When surface reservoirs of water begin to evaporate, the pollution in the water concentrates, making treatment more challenging. So does the flooding of reservoirs, rivers, and more, which can be overrun with excess farming chemicals and wastes and other pollutants that can wash into them.

Turbidity and color are optical properties of water that can vary greatly in surface supplies, groundwaters under the influence of surface water, and some regional groundwaters. Color can be inorganic or organic in nature, and its origin will dictate the type of treatment necessary. Color in surface water is derived from decomposed leaves and roots, other plant remains, and perhaps iron and manganese. Sewage can also be a source of color and is a health concern.

The pH of water, which is important to the treatment process, generally ranges from 6.5 to 8.5 standard units in surface waters but can vary from these values if pollution or acid mine runoff is present. Source water that varies too far from neutrality (pH = 7) will require some adjustment before distribution to customers. Groundwater pH generally ranges from 6.0 to 8.5. Hard water (see Chapter 9) contains calcium (Ca) and/or magnesium (Mg) and can cause problems in plumbing systems, often requiring a utility to incorporate softening into its treatment scheme (see Chapter 9). Table 4-2 shows hardness concentrations and the typical designations used to describe the water.

Organic constituents in water are a concern because they can be sources of disinfection byproducts (DBPs), as discussed in Chapters 1 and 2. They are also a source of taste-and-odor problems. Estimates from various sources show that there may be more than 1,000 organic compounds that have been identified in drinking water supplies.

Process Design

Pretreatment processes, usually found downstream of the source, are designed to condition the water before it enters the treatment plant. A well-designed and well-operated pretreatment system will save the treatment plant capital and operating expenditures. Often, such a system can remove or minimize contaminants that occur only occasionally and that otherwise might require additional in-plant treatment capabilities that are rarely used. It is almost always better to prevent source water contamination than it is to bring it into the plant for treatment.

Table 4-2 Hardness concentrations with typical designations

Common Designation	Hardness *mg/L as $CaCO_3$*
Soft	0–75
Moderately hard	75–150
Hard	150–300
Very hard	>300

$CaCO_3$—calcium carbonate
Source: AWWA 2010

Screening

Screens and microstrainers are often used in the pretreatment process to remove gross debris and macrobiota (fishes, zebra mussels) before they can enter the treatment plant and damage pumps and appurtenances. As such, these are primarily physical processes.

Fixed screens, or bar screens, made from timbers or metal bars, may be located at the submerged intake. Screen openings are 1–3 in. (2.5–7.5 cm) and are designed to allow for velocities of less than 2 ft/s (0.6 m/s). Screens should be cleaned using water jets or air-scour mechanisms. Some screens are movable and can be raised to the surface for manual cleaning. Traveling screens (Figure 4-1), which are used at larger facilities, have multiple screens attached to a movable pulley. The pulley moves continuously, always bringing a new and cleaned portion of screen to the intake. Clogged screens move to the surface and are jetted. Microstrainers are often positioned downstream of intake screens or on source supplies where small particles and algae are abundant.

Source: Al-Kamil/Shutterstock.com

Figure 4-1 Traveling screens at source water station

Oxidation

Source waters may be oxidated and aerated when the need arises; this need may be periodic or continuous. Iron and manganese, certain odors, carbon dioxide (CO_2)–laden waters, and water devoid of oxygen can be improved with preoxidation before the water comes into the plant. Refer to these chapters for specific information on oxidation process operation:

- Aeration: Chapter 10
- Chlorine (Cl), chlorine dioxide (ClO_2), ozone (O_3), ultraviolet (UV) light: Chapter 8
- Potassium permanganate: Chapter 10

Aeration is often the treatment of choice for iron removal. The oxidation state of iron can be elevated to +3, and this form combines with oxygen to become insoluble. The iron solids then settle or can be filtered more easily. Aeration is a very cost-efficient process for this purpose.

Manganese usually requires a more powerful oxidant to achieve conversion to the +4 form. This can be accomplished with free chlorine, chlorine dioxide, or ozone. Chlorine dioxide is an attractive choice because it does not form regulated organic DBPs. The dosage must be controlled, however, because chlorite (a byproduct) is regulated. About 2.5 mg/L of chlorine dioxide is required to remove 1.0 mg/L of manganese (site-specific conditions may alter this ratio).

Zebra and quagga mussels can be effectively controlled by the intermittent or continuous application of chemical oxidants. These small mollusks are invasive species that have caused serious problems in water supply pipelines. The organisms attach to the inside of transmission mains, where they can disrupt flow and cause expensive increases in pumping requirements. Some water systems have applied chlorine, potassium permanganate ($KMnO_4$), or chlorine dioxide at water intakes to control or eliminate these pests. Other methods of control include physical removal and UV disinfection near the point of water withdrawal.

Powdered Activated Carbon

To treat taste and odor or color in a water supply or to deal with algal toxins, it is sometimes useful to apply powdered activated carbon (PAC). The point of addition can be inside the WTP, usually before sedimentation, but can also be near the supply source. Applying PAC near the source may have advantages because more time is available for adsorption of the substances of interest. This is especially true if presedimentation basins are available. In many instances, PAC requires substantial contact time to work effectively. Also, the dosage needed may exceed the amount that could be applied in the treatment plant. PAC feed systems are similar to other dry chemical operations that include storage (Figure 4-2), dry feeders, slurry makeup, and solution feed components (see Chapter 5).

Source: Aqua Ohio–Struthers Division

Figure 4-2 Bulk storage silo with powdered activated carbon

Presettling

Operation of presedimentation basins usually requires regular removal of sand and grit; a sample program for testing the influent and effluent basins is a good idea. The frequency of testing can vary from season to season and can include an increase in sampling when water that is high in solids is taken in. As deposits accumulate in basins, the detention time lessens and more grit and residue enter the treatment train, where damage can occur. Anaerobic conditions can also develop, causing treatment problems. Some packaged pretreatment units are cleaned automatically, whereas large, uncovered reservoirs require periodic manual cleaning and drying.

Record-keeping practices are useful. The types and amounts of sedimentation, both influent and effluent, give a reliable indication of the unit's efficiency and can indicate a need for cleaning. Macrobiological sampling and analysis are also recommended, especially where zebra mussels and clams are present. The control of pests (e.g., flies, gnats, or aquatic weeds) may require special pretreatment techniques such as preoxidation or physical scraping.

BIBLIOGRAPHY

AWWA. 2010 (4th ed.). *Principles and Practices of Water Supply Operations—Water Treatment*. Denver: AWWA.

AWWA. 2004 (3rd ed.). M7, *Problem Organisms in Water: Identification and Treatment*. Denver: AWWA.

Chapter 5

Coagulation and Flocculation

Most surface water supplies, and many groundwater supplies, contain particles that are too small to be removed by sedimentation without the use of chemicals (see Chapter 2). Large, heavier particles may be removed in presedimentation basins; however, small particles can take weeks or months to settle. Smaller suspended particles tend to resist settling because of their size and their natural repellant forces. These natural forces are excess electrical charges that tend to keep the particles in suspension by helping the particles repel each other. This small negative electrical charge, which creates a repelling force much like the similar ends of a magnet, is called *zeta potential.*

Small nonsettleable solids are referred to as *suspended solids, colloidal solids,* or *dissolved solids,* depending on their chemistry, size, and origin. Suspended particles can be biological organisms, bacteria, viruses, protozoa, organic material, or inorganic solids. Together, they give the water a cloudy appearance that is called *turbidity.* Visible turbidity is objectionable to consumers. It can also be accompanied by harmful biological agents and therefore must be removed. Turbidity can prevent pathogens from coming in contact with the applied disinfectant. The Safe Drinking Water Act (SDWA) requires that turbidity be removed to very low levels (see Chapter 1) before distribution of water to customers.

To do this, water treatment plants (WTPs) normally apply chemicals to accelerate the removal of particles. These chemicals neutralize the zeta potential and allow the particles to bind together, become heavier, and settle. This process, known as *conventional treatment,* includes common concepts that operators should understand. Conventional treatment processes

- are designed to be operated at certain predetermined rates and under certain conditions;
- can be optimized for performance (and so it follows that they can be compromised and overwhelmed); and
- taken together, form multiple barriers against the passage of pathogens into the distribution system.

The unit processes that are primarily responsible for conventional-treatment particle removal include coagulation, flocculation, sedimentation, and filtration. The proper operation of these four processes will result in maximum removal of particulates from the source water and will produce a finished water with low turbidity. This chapter focuses on coagulation (rapid mixing here is considered as part of the coagulation step) and flocculation. Sedimentation and filtration are discussed in subsequent chapters.

COAGULATION

Coagulation is the addition and rapid mixing of chemicals, called *coagulants* or *coagulant aids*, in source water to destabilize colloidal particles.

Chemical Coagulants

Common coagulants used in conventional water treatment include alum, ferric chloride, ferric sulfate, sodium aluminate, and various cationic polymers. The polymers are either used alone or in a form that has been combined (blended) with aluminum (Al) or iron (Fe) salts (ferric salts). Both alum and ferric salts work by reacting with the water's natural alkalinity to form jelly-like particles of aluminum hydroxide or iron hydroxide called *floc*. Enough alkalinity must be present for this to occur. If there is not enough alkalinity for the reaction to take place, operators may augment the naturally occurring alkalinity by adding caustic soda, soda ash, or lime (quick or hydrated).

Aluminum and iron coagulants put a large amount of positively charged ions into the water. These metal salt coagulants consume alkalinity, causing the pH of the source water to drop. The resulting ions form positively charged flocs that begin to attract the negatively charged particles of color and turbidity. As the particles collide in the mixing area, they stick together and form increasingly larger particles.

The alkalinity of the source water often dictates whether the plant uses a "charge neutralization" approach or a "sweep floc" approach to flocculation. In very low-alkalinity waters, WTPs are often designed and operated as charge neutralization—i.e., alum or ferric is added in just enough quantity to neutralize the electrostatic charges. Many direct-filtration plants follow this route. It consumes very little alkalinity and produces relatively little residuals, so the sedimentation process is not needed.

Where source water alkalinity is higher (greater than 50 mg/L) and space for residuals handling is not an issue, operators mostly use the sweep floc approach, in which much more coagulant is added to create larger flocs that "sweep up" the smaller particles during flocculation. This approach is much more forgiving—a few milligrams per liter of coagulant lower or higher than the ideal dosage doesn't make as big a difference. It will create more sludge, but it is also more likely to remove particulate organic carbon.

One note of caution: In some older and larger plants, the operators must use “carry water” to get the coagulant from the day tank over long distances to the feed point. This has the effect of raising the pH of the coagulant, which makes it less effective. Alum should never be diluted by carry water to a pH higher than 3.3, and ferric salts solutions should never have a pH higher than 2.2.

The reactions of alum and ferric salts in water are governed by conditions such as mixing intensity, pH, temperature, alkalinity, and the nature and amount of turbidity. Alum works best in a pH range of 5.8–8.5, while ferric salts are effective over a wider range (4–11 in some circumstances). Both consume alkalinity as they form floc, causing the water’s pH to drop. Ferric salts form a denser floc than does alum, and both form floc at a slower rate as the source water temperature decreases. Operators usually compensate for this by increasing the coagulant dosage. Initial application of the chemical is critical to floc formation, making it necessary to position the point of contact where mixing intensity is highest. Water that is highly colored with organic material requires a lower pH for coagulation; therefore, acid addition before coagulation may be necessary.

COAGULANT AIDS

Coagulant aids are used to improve the coagulation process and build stronger, more easily settled floc. They can also help to compensate for low temperatures and lower the required amount of primary coagulant. Coagulant aids may be used to raise alkalinity or increase the amount of solids in the water. The wastes produced during the coagulation process are called *residuals*, and they must be accounted for and disposed of properly. The residuals are a function of the amount of turbidity in the source water and the amount of coagulant and coagulant aid used to remove it. Refer to Chapter 6 for more discussion of residuals.

Common Coagulant Aids

Activated Silica

Activated silica is a natural polymer that increases the rate of coagulation for alum and allows it to work over a wider pH range. It is typically used in amounts from 7 to 11% of the alum dose. The disadvantage of this aid is that the operator must activate it on-site. If the activation process is not carefully controlled, the silica will actually inhibit the floc formation and possibly clog filters.

Weighting Agents

If source waters are low in turbidity, it may be necessary to add particles to the water to enhance floc buildup. Weighting agents, such as bentonite clay, add particles to the water, which increases the probability of collisions between particles. Typically, dosages of 10–50 mg/L are used. This will add significantly to the amount of residuals produced.

Polyelectrolytes (Polymers)

Polymers are very large molecules that carry highly charged ionic sites. Depending on the balance of the electrical charges they possess, they are classified as cationic, anionic, or nonionic.

Cationic polymers possess an excess of positive charges and therefore are widely used to coagulate source waters with negatively charged particles. They are used as primary coagulants or as aids. Many WTPs use coagulants that are a prepared blend of alum or ferric salts with polymer. These blends help reduce the amount of metallic salts needed and therefore reduce pH depression. They also tend to form a tougher, more rapidly settling floc. They are very effective in flocculating living organisms such as algae. Cationic polymers are also commonly applied in very dilute solutions directly to waters just before filtration to improve filtration performance (filter aids).

Anionic polymers possess an excess of negatively charged sites and are used for source waters or other process points in the plant where turbidity is positively charged. These compounds are often used at very low dosages as filter aids.

Nonionic polymers, as the name implies, possess a balance of anionic and cationic sites and therefore have an overall neutral charge. They release both positive and negative charges into the water and can coagulate waters with varying zeta potentials. Low dosages are often used when these compounds are applied directly ahead of filtration (filter aids).

Anionic and nonionic polymers are typically suited as flocculation aids (fed just after cationic flocs are formed, e.g., after a minute or so of mixing) or for residual thickeners (in solids settling tanks or before application to a sludge press, sand bed, or centrifuge). These polymers are often formulated from regulated substances (acrylamide or epichlorohydrin) and are usually fed at low doses (less than 1 mg/L). Their use in drinking water treatment requires verification that they will be maintained at safe levels. Overdosing of polymers can lead to poor sedimentation and filtration.

Coagulation Chemical Feed Systems

Successful coagulation depends not only on the choice of coagulant but also on reliable feed equipment and proper mixing facilities. *Additionally, operators must have a basic understanding of dosage control and process optimization.*

Dosage Control and Process Optimization

An operator in a WTP uses the plant chemicals and processes to render source water into drinking water and so is one of the most important duties that an operator has while on shift. An operator should never rely

Source: Lake County, Ohio, Department of Utilities

Figure 5-1 Alum feeders at Aquarius WTP

on supervisory control and data acquisition (SCADA) to do this without first knowing what to expect. For volumetric feeders and small-dose gravimetric feeders, this means that the operator must do four things over and over for each of these chemicals fed:

1. Decide on a dosage for the chemical, either by a lab bench test or experience.
2. Set the feeder to deliver the amount of chemical per unit of time needed to achieve the dosage.
3. Check the feeder to determine if the chemical is actually feeding that amount of chemical; if not, adjust it.
4. Test the processed water to determine if the chemical is achieving the desired result; if not, adjust the dose and go back to step 2.

An example for this determination might look like this:

The operator at the WTP intends to treat the average 2023 daily flow of 5.14 mgd with 22-mg/L dry-basis alum. How many milliliters per minute must be fed?

1. Convert the flow of 5.14 mgd to liters per minute:
   - *5,140,000 gal/d × 1 day/1,440 min × 3.785 L/gal = 13,510 L/min*
2. Calculate the amount of dry-basis alum per minute needed:
   - *22-mg/L dry-basis alum × 13,510 L/min = 297,220-mg/min dry-basis alum*
3. Calculate the milliliters of liquid alum per minute needed:
   - *297,220-mg/min dry-basis alum fed × 1 mL/642 mg = 462-mL/min liquid needed*

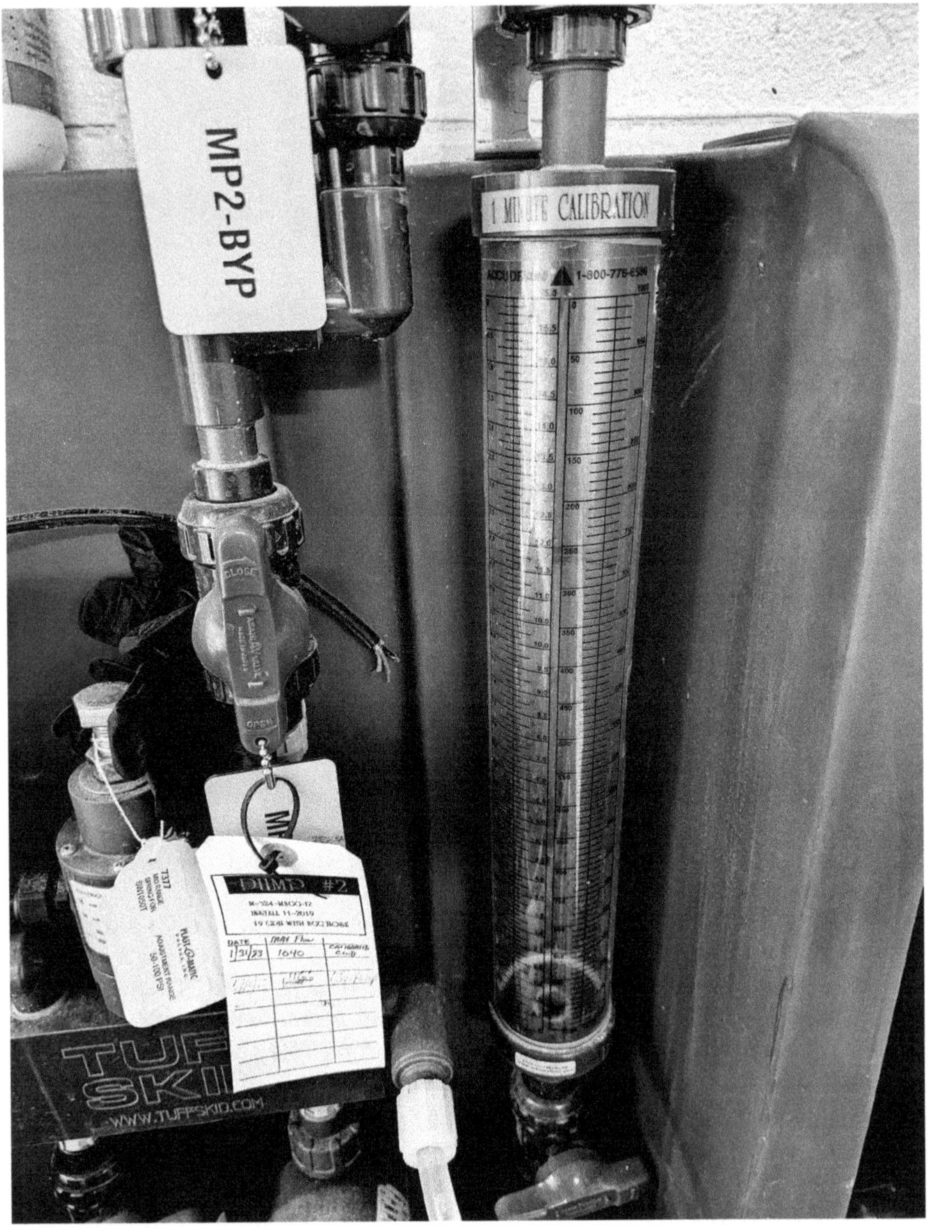

Source: Aqua Ohio–Struthers Division

Figure 5-2 Calibration column on volumetric feeder

The operator now knows how many milliliters per minute of liquid alum to feed, so the machinery must be set to feed that amount and *checked for accuracy*. The feed equipment should have a calibration column attached that allows the feed pump to draw liquid chemical from it, while the day tank is isolated. In this way, the operator can view the amount fed in 1 min.

The procedure is to fill the calibration column at the feeder with liquid alum and turn the valves so that the feed must come only from the column. The column should be sized so that it can hold more than is needed in 1 min's time. Using a stopwatch, the operator watches the liquid begin to fall in the column as it is being fed—preferably for two or three feed volumes.

> *Example*: Say the operator is expecting to feed about 117 mL/min. A ½-L column would allow the operator enough volume to time the feed for 3 min. After 3 min, the operator takes the reading, which should be close to (3 × 117) = 351 mL fed.

Chemicals are stored as liquids or solids, depending on the amounts used and the size of the plant. Larger plants tend to use liquid products, which can be delivered in bulk. Many of the polymers that are used in smaller amounts can be handled in dry form, which is delivered in bags or drums. Operators usually prefer liquid chemical storage and handling because it creates less dust, fewer safety concerns, and fewer inventory problems. The choice of chemical type will have a bearing on the type of feed system used.

> *Again: The operator should not depend on SCADA to do these tasks for him or her.*

Dry Chemical Feed Systems

Two general types of dry chemical feeders are used in WTPs: volumetric and gravimetric. Both systems are equipped with mixers and a batch or mixing tank that provides a place for the chemical to mix with some carry water. This disperses the chemicals into solution or suspension before the application point. It is important that dry chemicals be thoroughly mixed before they are added to the treatment process.

Volumetric feeders (Figure 5-3) transfer a chosen volume of chemical from a hopper to a mixing tank in a unit of time. They have adjustable settings that allow for dosage control. Volumetric feeders are not as accurate as gravimetric feeders but are less costly. Gravimetric feeders (Figure 5-4) measure out a predetermined weight of chemical per unit of time.

Both volumetric and gravimetric systems can be operated automatically or manually. When operated in the automatic mode, they are often set to run from a signal of source water flow or zeta potential or from other such controllers. Operators should understand that these automatic systems can fail and therefore do not guarantee trouble-free operation. Feeders should be checked systematically to confirm that they are doing what they have been programmed to do. Automatic controls can, however, provide a measure of reliability and free operators to perform other tasks. Also, automatically controlled feed systems allow for dependable operation of treatment plants from remote locations.

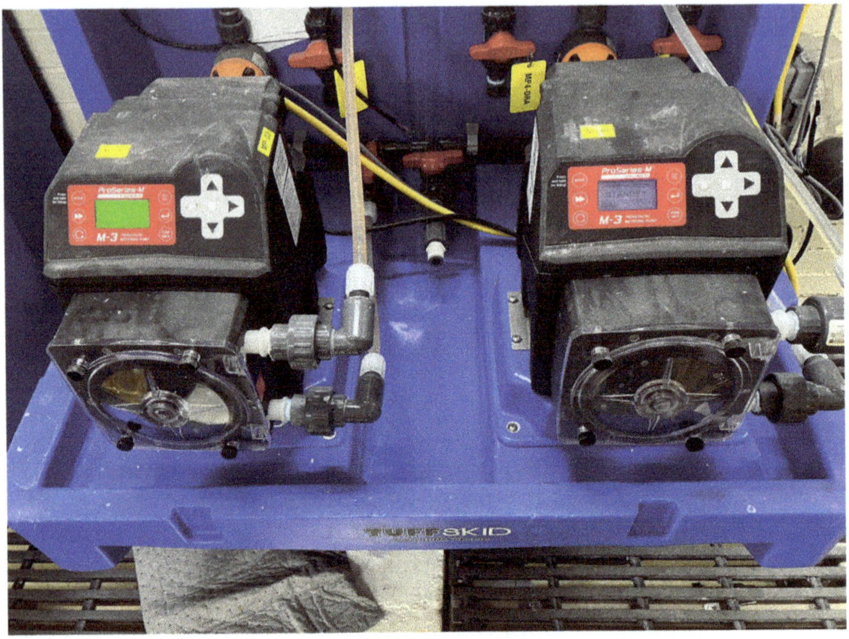

Source: Nick Pizzi

Figure 5-3 Volumetric feeder: Feed pumps

Source: Aqua Ohio–Ashtabula

Figure 5-4 Gravimetric feeder

Solution Feed Systems

Liquid coagulants and coagulant aids are fed either full strength or as a dilution using solution feeders. When the chemicals are fed "neat" (directly from the package without dilution), they are commonly delivered through a metering pump. These positive-displacement units are usually variable-speed driven and can be paced from a signal of choice (e.g., source water flow or zeta potential). Small solution feed systems offer the advantage of low maintenance and excellent, precise control and can be set up to feed directly from chemical drums that are delivered to the plant. For very small applications, peristaltic pumps are often used. These pumps have a roller mechanism that squeezes the coagulant through a tube or flexible hose, which must be replaced periodically. They are very accurate, even when used in larger facilities. Also, because no chemical actually touches the pump, there is no internal corrosion problem.

Solution feeders can be arranged with several tanks in series, one of which can be used as a dilution tank. With this setup, a higher-strength chemical can be purchased and dilutions made and fed as needed. This allows for less frequent purchases and perhaps a lower unit price for chemicals.

For chemical inventory control, a solution feed tank can be set on a scale to accurately measure weight loss, or the tank can be calibrated with markings at intervals so that volume-loss measurements can be recorded. Alternatively, the tanks can be equipped with sonic measurement devices that give accurate readouts of tank volume.

Expressing Dosage

When talking about their dosage, operators must understand the difference between "dosage as product" and "dosage as ingredient." As AWWA Manual M37, *Operational Control of Coagulation and Filtration Processes*, explains, one way to avoid confusion is to express dosages and concentrations in terms of the molar amount of active ingredient—i.e., the moles of iron or aluminum per liter of solution. This concept has an even greater importance for jar tests when comparing different coagulants. Equal molar dosages of aluminum and iron contain the same number of atoms. To determine the molar metal dosage applied at the full-scale treatment plant, data for the plant flow rate, data for chemical feed rate, and the chemical data sheets with specific gravities are needed. Data sheets that are usually provided for alum and ferric will provide the percent active ingredient, which is percent aluminum oxide (Al_2O_3) for alum and percent Fe for ferric. To convert to molar metal dosage, use this formula:

$$\text{Dosage, micromoles/liter} = \frac{\textit{Micromoles per Liter}}{\textit{Plant Flow Rate} \times 8.34} \times \text{Factor}$$

The "factor" is provided in Table 5-1 for several of the more common coagulants used.

Table 5-1 Factor needed to convert molar metal dosage, based on coagulant used

Coagulant	Factor
Dry alum	3.365
Liquid aluminum product (alum, polyaluminum chloride [PACl])	% Al_2O_3 by weight ÷ 5.1
Dry ferric chloride	6.165
Dry ferric sulfate	5.004
Dry ferrous sulfate	6.579
Liquid iron product (ferric sulfate, polyferric chloride)	% Fe by weight ÷ 5.58

Source: AWWA 2011a

AWWA M37 provides an example, which is shown as this question: A WTP feeds 5,000-lb/d liquid alum and treats a flow rate of 5 mgd. The alum is 8.3% Al_2O_3 by weight. What is the molar equivalent dosage?

Example in AWWA M37	Traditional Operator Calculation
Dose: µmoles per liter	*Dose*: µmoles per liter
5,000/(5 × 8.34) × (8.3/5.1) = 195 µmol as Al	5,000 lb of liquid alum × 1 gal/11 lb × 3.785 L/gal
AWWA M37 then uses a chart and states that 195 µmol as Al is equal to 58-mg/L alum.	5,000,000 gal/d × 3.785 L/gal = 1.89 × 10^7L/d We know that commercial alum is 642-g/L dry-basis alum. (1,720.3 L × 642 g/L)/(1.89 × 10^7 L) = 0.058 g/L, or *58-mg/L alum*

The chart-style graphic provided by AWWA M37 (Figure 5-5) shows how to convert the 195 µmol/L to 58 mg/L. It also provides a convenient way to compare aluminum and ferric coagulants on a mole-per-mole basis.

This may be required of some utilities that cannot meet the requirements of the coagulation end point of the Disinfectants and Disinfection Byproducts Rule, which states that if you cannot meet the total organic carbon (TOC) removal requirements or the six "alternative criteria," then you must meet your "enhanced coagulation endpoint." This is a pH end point that is met when TOC is not reduced more than 0.3 mg/L at incremental dosages of ***10-mg/L dry-basis alum or equivalent Fe.***

Table 5-2 shows the lowest coagulation-zone pH that must be met if a utility needs to attain the enhanced coagulation end point, which is the combination of coagulant, or coagulant with acid addition, that produces the pH needed. The pH can be determined by jar testing rather than by a full-scale demonstration.

Table 5-2 Lowest coagulation-zone pH needed to attain enhanced coagulation

Alkalinity *mg/L*	pH Value
0–60	5.5
>60–120	6.3
>120–240	7.0

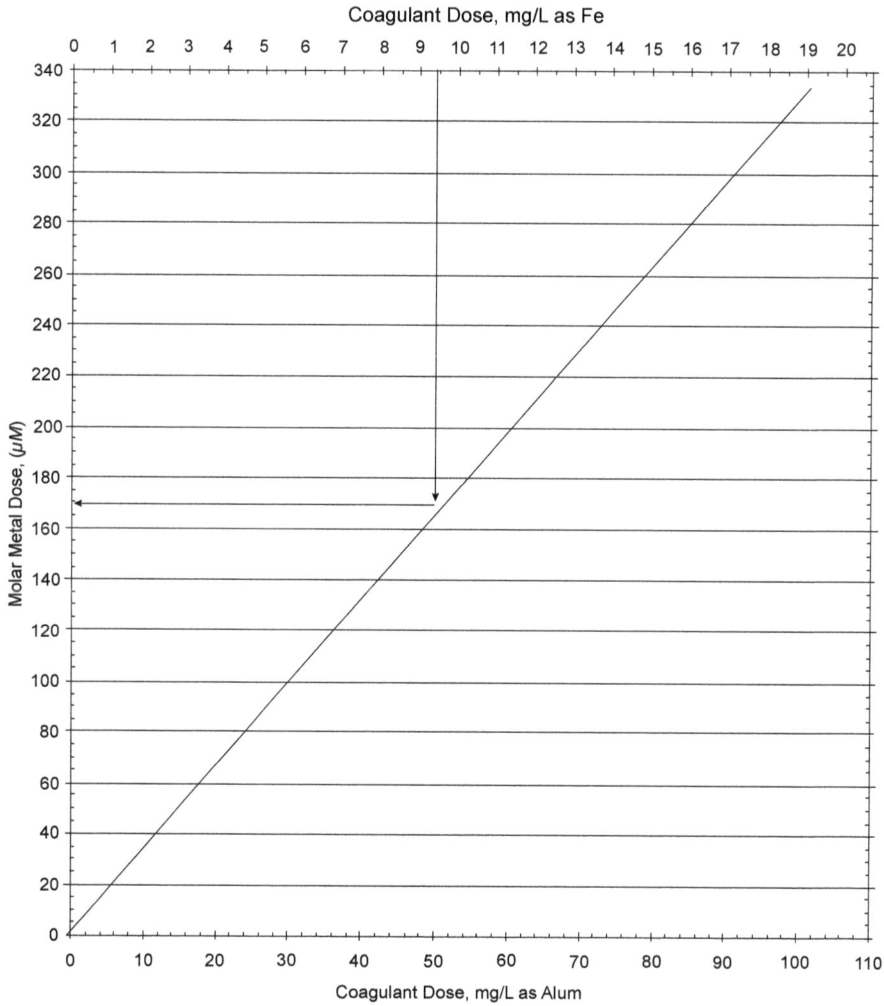

Source: AWWA 2011a

Figure 5-5 Molar metal equivalent chart

PROCESS OPERATIONS AND MAINTENANCE

Process control instrumentation, such as streaming current monitors, can help operators to produce a quality product. To achieve high-quality water, a WTP must be operated in a consistently optimized manner, and plant personnel should be able to

- follow a written procedure (standard operating procedure) that outlines and directs bench- or pilot-scale testing (jar testing, etc.) that determines proper chemical application dosages and sequences for their plant and
- apply the knowledge gained in this determination to the plant at full scale.

The details of how to perform jar tests are described in Chapter 12 and in AWWA M37.

Alum, the most widely used chemical in the coagulation process, has the formula $Al_2(SO_4)_3 \bullet 14.3H_2O$. As dry alum, it is 17% basis Al_2O_3. Each milligram per liter of alum added to water will consume approximately 0.5-mg/L alkalinity. Liquid alum, purchased and used in bulk quantities, is actually dry alum diluted with water to about 48.8%, dry basis. Liquid alum weighs approximately 11 lb/gal at an average specific gravity of 1.325 (8.34 × 1.325). Liquid alum contains about 5.36 lb/gal dry-basis alum (48.8% of 11 lb/gal). With this information, the strength of alum, in milligrams per milliliter, can be calculated and used in dosage problems. The calculation is:

$$5.36\text{-lb / gal Dry Alum} \times 1\text{ gal / }3.785\text{ L} \times 453.6\text{ g / lb} \cong 642\text{ g / L, or }642\text{ mg / mL}$$

Therefore, each milliliter of liquid alum puts 642 mg of dry-basis alum into the water. Dosage problems can now be solved, as in the following example.

Example: After performing a jar test, a WTP operator chooses a dosage of 24-mg/L dry-basis alum. The plant is operating at 1.6 mgd. How many milliliters per minute of liquid alum must be fed?

1. Convert 1.6 mgd to liters per minute.

$$1{,}600{,}000\text{ gal/d} \times 3.785\text{ L/gal} \times 1\text{ day/}1{,}440\text{ min} = 4{,}205.5\text{-L/min Water Treated}$$

2. Calculate the amount of liquid alum needed.

$$24\text{ mg/L} \times 4{,}205.5\text{ L/min} \times \text{mL of Alum/}642\text{ mg} \cong 160\text{ mL/min}$$

Alternative check on calculation:

$$160\text{ mL/min} \times 1{,}440\text{ min/d} \times 642\text{ mg/mL} \times \text{lb/}453.6\text{ g} = 326\text{-lb/d Dry Alum}$$

$$326\text{ lb/d Dry Alum} \div 1.6\text{ mgd} \div 8.34 = 24\text{ mg/L}$$

After performing the calculation and setting the feed equipment to deliver the needed amount of chemical, the operator must periodically check the equipment to determine if it is actually feeding the required amount. For liquid feed systems (described in the previous example), operators must have a way to capture or measure the amount of liquid fed per unit time, usually each hour. Dry feed systems require a method for measuring the weight of chemical fed per unit time. Both systems are normally equipped with such features. Liquid feed systems that use metering pumps are sometimes rigged with volumetric site tubes on the suction side of the pumps. Older volumetric feeders, such as the "rotodip" feeder, require the operator to capture the liquid as it is fed or to know the volume displaced from each dip. A chart for feed rates of liquid alum is given in Table 5-3.

For source water pump rates other than those shown, use the appropriate multiplier or extrapolate. Two examples follow:

> *Example 1:* A WTP operator is making a pump change that will create a total source water flow of 2.5 mgd. The desired alum dosage is 20 mg/L. How many milliliters per minute of alum should be fed?
>
> *Answer:* A 20-mg/L dose for 2 mgd requires 164 mL/min. A 20-mg/L dose for 3 mgd requires 246 mL/min. Extrapolating between the two shows that about 205 mL/min is needed.
>
> *Example 2:* A jar test indicates that 17 mg/L of alum should be fed into a source water flow of 80 mgd. How much alum (in milliliters per minute) is needed?
>
> *Answer:* To treat 8 mgd with a 17-mg/L dose of alum, 557 mL/min is needed. Therefore, to treat 10 times as much flow, or 80 mgd, multiply 557 by 10. Thus, 5,570 mL/min is needed.

Some WTPs use an alum–polymer blend as their primary coagulant. Normally, the dosage for this product is expressed as gallons per million gallons, or parts per million. This is based on the simple idea that 1 gal of product added to each million gallons of source water is 1 ppm. This expression negates the requirement to know the exact proportions of alum and polymer in the coagulant.

Note that parts per million and milligrams per liter are not necessarily the same. Parts per million is a volume-per-volume relationship, while milligrams per liter is a mass-per-volume relationship. As a rule of thumb, a dosage of 18 ppm would deliver approximately 10.6-mg/L dry-basis alum. The calculation is as follows: Feeding 18 ppm of alum–polymer blend into the source water stream is the same as feeding 18 gal of product per 1,000,000 gal of water. An alum–polymer blend weighs about 10.9 lb/gal (it may vary), so 18 gal × 10.9 lb/gal = 196.2 lb/mil gal and 196.2 lb/mil gal/8.34 = 23.5-mg/L liquid alum–polymer blend. Because roughly 45% of this is dry alum, 45% × 23.5 = 10.6-mg/L dry-basis alum. Because more dry-basis alum is usually required to remove TOC than is needed to remove turbidity, operators should resist lowering blended dosages.

Table 5-3 Amount of liquid alum needed to achieve a desired dry-basis dosage at a given pumpage

Dry-Basis Alum Desired *mg/L*	Instantaneous Flow Rate Being Treated *mgd* 1	2	3	4	5	6	7	8	9
	mL/min **Needed**	*mL/min* **Needed**	*mL/min* **Needed**	*mL/min* **Needed**	*mL/min* **Needed**	*mL/min* **Needed**	*mL/min* **Needed**	*mL/min* **Needed**	*mL/min* **Needed**
12	49	98	147	196	246	295	344	393	442
13	53	106	160	213	266	319	372	426	479
14	57	115	172	229	286	344	401	458	516
15	61	123	184	246	307	368	430	491	552
16	65	131	196	262	327	393	458	524	589
17	70	139	209	278	348	417	487	557	626
18	74	147	221	295	368	442	516	589	663
19	78	155	233	311	389	466	544	622	700
20	82	164	246	327	409	491	573	655	737
21	86	172	258	344	430	516	602	687	773
22	90	180	270	360	450	540	630	720	810
23	94	188	282	376	471	565	659	753	847
24	98	196	295	393	491	589	687	786	884
25	102	205	307	409	512	614	716	818	921

Source: Nick Pizzi

NOTE: Table assumes that liquid alum is 642 mg/mL dry-basis alum.

To use the table, follow these steps:

1. Determine, through jar testing or some other means, the desired dosage of dry-basis alum.
2. Locate that dosage in the left column of the table.
3. Choose a treated water flow rate in the top row of the table.

The intersection of the column and the row shows the approximate amount of liquid alum needed, in milliliters per minute, to achieve the desired dosage.

Table 5-4 Amount of alum–polymer blend needed to achieve various dosages at several source water flow rates

Desired ppm	20 mgd mL/min (gal/d)	40 mgd mL/min (gal/d)	60 mgd mL/min (gal/d)	80 mgd mL/min (gal/d)	100 mgd mL/min (gal/d)	120 mgd mL/min (gal/d)	140 mgd mL/min (gal/d)
15	789 (300)	1,578 (600)	2,367 (900)	3,156 (1,200)	3,945 (1,500)	4,734 (1,800)	5,523 (2,100)
18	947 (360)	1,894 (720)	2,840 (1,080)	3,787 (1,440)	4,734 (1,800)	5,680 (2,160)	6,628 (2,520)
21	1,105 (420)	2,210 (840)	3,314 (1,260)	5,302 (1,680)	5,523 (2,100)	6,628 (2,520)	7,732 (2,940)
24	1,262 (480)	2,525 (960)	3,788 (1,440)	5,050 (1,920)	6,312 (2,400)	7,574 (2,880)	8,837 (3,360)

Source: Nick Pizzi

Table 5-4 shows feed amounts of alum–polymer blends needed, in milliliters per minute and gallons per day, to achieve various parts-per-million dosages.

As discussed earlier, iron-based coagulants, such as ferric chloride and ferric sulfate, are also used in water treatment. In some cases, they may be a better coagulant choice because of their ability to work in a wider range of pH. Because iron salts tend to consume more alkalinity than do aluminum salts, they may need less mixing or less contact time for reactions to take place but may depress the pH further. The difficulty that is most often cited when using ferric and ferrous coagulants is their corrosiveness to the storage vessels and feed systems, which affects the maintenance efforts of operators.

Operational Unit Processes

In the conventional WTP, the primary job of an operator is to obtain source water and use the chemicals and the physical tools provided to them to render the water fit for human consumption. That fitness is determined by adherence to the SDWA regulations. The physical tools that the operator uses are called "unit processes." A unit process in a WTP is an engineered component that has a specific purpose in the overall treatment scheme:

1. It is an important part of the overall function of the WTP.
2. It is meant to be operated—it is operable.
3. It has well-defined operating parameters: Its success as a process depends on it being operated within those parameters.
4. Its performance or results can be measured because it will have input and output data associated with it.

Engineers and regulators follow agreed-upon standards when designing, approving, and building major unit processes. Operators can study these standards as a means of understanding how each can be used in an optimized fashion to help meet regulations and goals for finished water. Operators are ultimately responsible for using these unit processes as intended to produce the results they need.

One common source of unit process design requirements can be found in *Recommended Standards for Water Works*, also known as the Ten State Standards. Many state regulatory agencies follow these standards or publish their own, and it is important for operators to become familiar with them. The better the operator understands the purpose, capabilities, and limitations of the unit process, the more likely he or she will obtain good results.

For conventional treatment facilities, the major unit processes are rapid mixing, flocculation, sedimentation, filtration, and disinfection. Other processes, such as precipitative softening and membrane filtration, are covered in later chapters of this book.

Stabilization and Destabilization

In previous chapters, we saw that most particles in nature are colloids and carry a negative charge. Operators can think of these negatively charged particles as the ends of magnets. They will continually repel each other and stay suspended. They will do this for long periods of time because they exist in a stable state. The operator must add a chemical with enough positive charges in an amount roughly equal to cancel out the negative charges—or *destabilize*—the colloid. When this happens, the particles will settle quickly and filter well.

The book *Water Quality and Treatment* (AWWA 2011) states that "the object is to transform a stable suspension, that is, one that is resistant to aggregation or attachment to a filter grain, to an unstable one."

Mixing Facilities

Rapid Mix

Rapid mixing is an important part of the coagulation process, regardless of the coagulant used or the method for feeding the coagulant. Rapid mixing describes the intense mixing that usually takes place in the first moments when the coagulant is fed into the source water. It also describes the equipment that is used for the intense mixing period. The purpose of rapid mixing is to briefly and violently agitate the water and coagulant so that particles have numerous opportunities to collide. This process can easily fail in two ways: if the length of agitation is too short or too long or if the mixing energy is insufficient. To overcome these possibilities, engineers have devised several methods, described here, for rapid mixing.

Important Design Parameters for Rapid Mixing

G value. Successful rapid mixing requires that energy input to the water takes place at or just after the dosing point, and the amount of energy put into the water is stated as the velocity gradient, or *G* value. The *G* value should be sufficiently violent to thoroughly disperse the coagulant to every drop of water.

G value is typically stated as a number followed by the units of "seconds to the minus 1 power," e.g., 700 s^{-1}. Current standard engineering practice is to provide the WTP with rapid mixers that can develop *G* values in the range of 750–1,000 s^{-1}. At 30 seconds of detention time, the "$G \times T$" value is 21,000–30,000. Inline mixers must be much more violent because the detention time is very brief.

In the latter part of the 1970s to the early 1990s, it was not uncommon to see rapid mix designs provide smaller energy input units. For example, a *G* value calculation taken from the operations and maintenance manual of a 1982 WTP was shown as

$$G \text{ value} = \sqrt{(P_w / \mu Vt)}, \text{ where } P_w = k / g\ w\ N^3 D^5$$

- K = impeller constant 1
- w = 62.4 lb/ft^3
- N = revolutions per second (rps) at 233 rpm, so 3.88 rps
- D = impeller diameter 16.4 in., or 1.37 ft
- g = acceleration because of gravity, 32.2 ft/s^2
- $\mu = 3.76 \times 10^{-5}$ lb·s/ft^2 in cold water
- Vt for WTP mixer is 142 ft^3
- G calculates to a range of about 280–345 s^{-1}, so

$$G \times T \text{ is } (345 \times 30 \text{ s}) = 10{,}350$$

Mechanical mixers—propeller, impeller, or turbine type—can be placed in square or rectangular chambers. The detention time in these chambers is very short—usually 15–45 s. Most plants are equipped with multiple units so that operators can take them in and out of service. In this way, the range of desired detention times can be achieved. Operators should closely monitor these times and make every effort to maintain them. Units are normally equipped with gates or valves for ease of control. The chemical application point must be located at or very near the mixer blades to allow for quick dispersion of chemical.

Mechanical inline mixers have a motor and propeller mechanism placed directly into the pipe. Agitation must be very powerful to overcome the short detention time.

Static mixers provide good mixing by placing fixed vanes inside of the pipes carrying the source water flow. These units are economical and relatively free of maintenance requirements but do increase the head loss in the system.

Baffled chambers are sometimes used to provide mixing for chemical applications. They are limited in their ability to provide turbulence, which is controlled by the rate of source water flow.

Pumps used for feeding chemicals sometimes double as mixers. The coagulant is added to the suction side of the pump; the pump itself mixes the chemicals. Like the baffled chamber, there is little control over the amount of mixing.

Some typical design requirements for rapid mixers are

- detention time not greater than 30 s,
- provision for adding a primary coagulant at all times,
- mechanical mixing (static inline will be considered) capable of developing *G* value not less than 750,
- *G* value to be determined by jar testing, and
- mixers to be placed as near to the flocculators as possible.

Flocculation

After coagulation, the newly formed floc particles are brought into the flocculation stage, during which gentle mixing allows them to grow. This physical process is also governed by detention times and mixing energies. The process of particle size growth in flocculators is sometimes called *agglomeration*. Newly formed floc particles need gentle mixing for some time to provide many collision opportunities. As particles grow, traveling through the flocculators, they become more fragile and require less energy—the energy is tapered. But those same particles also grow heavier and must be kept in suspension so they can be carried over to sedimentation. The goal is to allow each discrete floc particle to undergo the flocculation process as close to the ideal detention time as possible. For this reason, flocculators are usually baffled and provided with a variable-speed drive unit (Figure 5-6).

Compartmentalization. The baffling creates compartments in the flocculators that reduce short-circuiting. Most modern flocculators are equipped with two or three sets of over-and-under baffles, which create three or four compartments. As water flows under and over the baffle walls, mechanical mixing is provided to each compartment. Each successive compartment receives less mixing energy than the previous compartment. It effectively keeps more of the water in the flocculator so that it can increase the possibility that any given floc particles will collide with its neighbors at or near the ideal detention time. This allows the floc particles to grow yet receive progressively gentler treatment.

The benefits of compartmentalization are shown in the following graphic taken from *Water Treatment Plant Design* (ASCE, AWWA, CSSE, 1997). The graphic, as shown in Figure 5-7, shows the ratio for actual versus theoretical retention time and helps explain how short-circuiting can impair flocculation, while the addition of compartments enhances it because more of the water is subject to the theoretical detention time as needed. It answers the question: "How much of my water is exiting the flocculator before it has had a chance to experience collisions as it should?"

We can answer that question using the graphic. For example, suppose we calculate the theoretical or average detention time of a basin using the formula *Dt* = Volume/Flow Rate. For a simple rectangular basin measuring

Source: Nick Pizzi

Figure 5-6 Variable-speed flocculator drive unit

16 ft wide and 55 ft long with a water depth of 16 ft and a flow rate of 3,500 gpm, the detention time would be:

$$Dt = \text{Vol} / \text{Flow Rate} = \frac{16\,ft \times 55\,ft \times 16\,ft \times 7.48\,gal\ per\ cubic\ foot}{3{,}500\,gpm} \approx 30\,\text{min}$$

We then know that ideally, if all of the water traveled through the basin as a unit, or as in "plug flow," then all of the particles in the water would receive ideal flocculation of 30 min—no more, no less. That would give us an "actual to theoretical" ratio of exactly 1, and we can locate that on the x-axis at 1.0.

However, water doesn't flow through pipes and basins as a plug; it is subject to eddies and currents and fluid drag along the side walls. This means that there is the tendency for some of the water to escape the basin sooner than the average time of 30 min and not have adequate time to agglomerate. This also means that some particles stay in the mixer longer than the ideal 30 min and tend to be broken up by too much mixing.

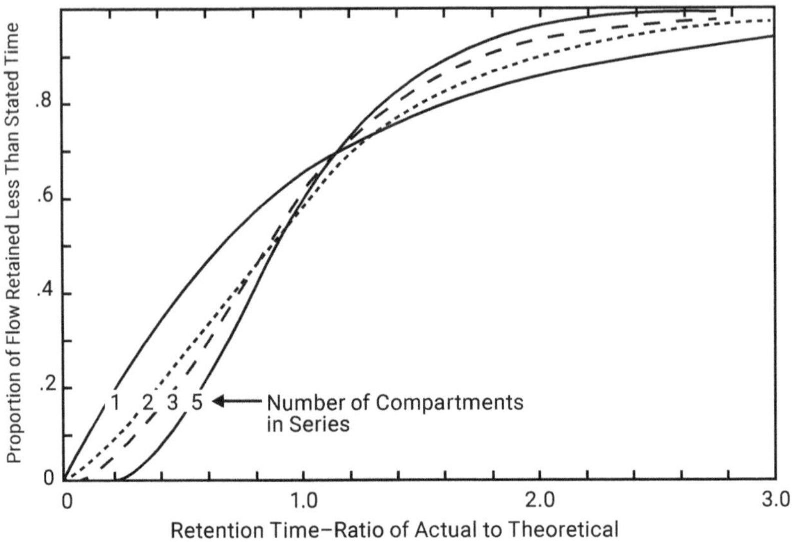

Source: AWWA 2011b

Figure 5-7 Graphic showing the ratio of actual versus theoretical detention time of a baffled flocculator

To minimize those tendencies, engineers design the baffles into the basin, which create compartments.

Referring back to our example of the 30-min ideal theoretical detention time and an actual to theoretical ratio of, say, 0.5:1 (or 15 min rather than 30) instead of 1:1, we can see that in as little as half that ideal time (15 min), a basin with one compartment (no baffle walls) will allow 42% of the water to escape without adequate flocculation time. But if that same basin is equipped with three compartments, that allowance reduces to about 25%, which is a huge improvement. That kind of information allows the operator an opportunity to optimize this important process. An operator cannot control the number of compartments that have been built into the flocculators but can control the number of units kept in service so that the theoretical detention time is met as closely as possible. This will serve to maximize the number of floc particles that receive the right amount of mixing.

Neglecting the important step often results in too little flocculation time. It can also create a situation in which some particles get too much detention time, such as when too many of the available flocculators are kept in service when the plant is operating at lower flow rates. This too is a hindrance to optimize treatment because the flocculators are perhaps overflocculating many of the particles and defeating the purpose of the process.

In commonsense terms, we can think of it this way: If a 20-mgd WTP has been designed with six equal-sized flocculation units, each rated at 3.33 mgd and 30 min of detention time, it doesn't make any sense to operate all of the flocculators when the plant is treating only, say, 8 mgd, because many of the particles are going to be in those flocculation units way too long. Three will probably do just fine in most cases.

Tapered energy. In addition to operating the flocculators in the ideal *Dt* range, the operator should also choose a tapered energy approach if the equipment is configured with variable drive mechanisms. Tapering the energy input allows for flocs to grow, while forces that would break them apart are continually lowered from one compartment to the next.

This is accomplished in several different ways. Many traditional flocculators are designed with a horizontal drive shaft that has a decreasing number of paddles mounted on it from one compartment to the next. For example, the shaft has paddles fixed to it: The first compartment may have four paddles; the next, three; and the final, two paddles. The shaft turns at the same speed, but less energy is imparted where there are fewer paddles.

Another traditional design is for each compartment to be equipped with dedicated vertical mixer and motor assemblies with variable-speed drive. Separate motors and shafts of the same size can be supplied for each compartment, and the size of the mixer blades is reduced. Each unit then can be "dialed in" to the desired energy output.

As discussed, flocculators are designed to provide 15–45 min of detention time, and a common operational mistake is to keep too many units in service, thereby providing excessive flocculation time. This can break up fragile flocs, creating a need for excessive amounts of coagulant. It is not uncommon to see flocculators with large amounts of settled floc buildup because the detention time was too long. This also creates a maintenance problem—units have to be cleaned more often. If they are not cleaned, sludge can turn septic or build up to the point of paddle impingement.

Tip speed of paddle in last compartment. In addition to operating the proper number of flocculators, operators must control the mixing energy imparted to the floc chambers. Control is designed into most modern flocculators so that an operator can choose the "tip speed" for the final stage. A good rule of thumb is that the tip speed of the final flocculator paddles should be about 1 ft/s and operate in the range of 0.9–1.3 ft/s most of the time (Figure 5-8). In

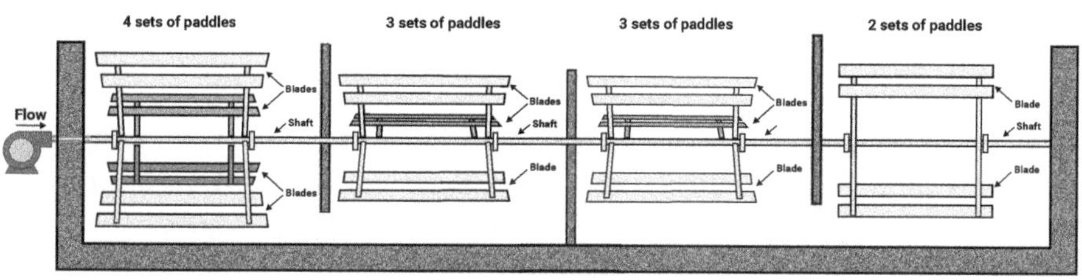

Source: Nick Pizzi

NOTE that there is only one shaft, but the section of shaft in each compartment has increasingly fewer or smaller paddle blades.

Figure 5-8 Drawing of a horizontal shaft flocculator with four compartments

colder water or when floc formation is heavy because of naturally occurring turbidity, a faster speed may be needed. The tip speed of the final paddle is determined using a stopwatch to measure the time it takes for the paddle to make one revolution. If the distance from the center of the shaft to the end of the paddle is measured, the distance traveled in that one revolution can be calculated as $\pi \times D$, where D is the diameter of the circle, in feet, that the outermost portion of the paddle makes when it revolves. D is calculated as two times the radius. The distance is divided by the time to obtain the tip speed, in feet per second. For example, if the final paddle in a flocculator revolves once every 55 s and the distance from the center of the shaft to the tip of the paddle (radius) is 11 ft, what is the tip speed?

Distance traveled by the paddle is $\pi \times D$, or
3.14 × 22 ft, which is 69 ft.

69 ft divided by 55 s is 1.25 ft/s.

Some typical design requirements for flocculation are as follows:

- Inlet and outlet design that minimizes short-circuiting
- Detention time of at least 30 min with a flow-through velocity of more than 0.5 and less than 1.5 ft/min
- Provisions to be made for tapered mixing energy
- Agitators driven by variable-speed systems capable of driving the peripheral tip of the paddles from 0.5 to 3.0 ft/s

Combined Units

Some process units combine rapid mixing, flocculation, and sedimentation in the same unit. Referred to as *solids-contact units* or *contact clarifiers*, these units are generally circular and contain equipment for mixing and sludge withdrawal. There is more discussion on these units in Chapter 6.

BIBLIOGRAPHY

AWWA. 2011a. M37, *Operational Control of Coagulation and Filtration Processes*. Denver: AWWA.

AWWA. 2011b (6th ed.). *Water Quality and Treatment* (J.K. Edzwald, editor). New York: McGraw-Hill.

AWWA. 2010 (4th ed.). *Principles and Practices of Water Supply Operations—Water Treatment*. Denver: AWWA.

AWWA. 1969. *Water Treatment Plant Design*. New York: McGraw-Hill.

Linder, K. and B. Martin. 2015. *Self-Assessment for Water Treatment Plant Optimization*. Denver: AWWA and Partnership for Safe Water.

Routt, J.C. 2001. Personal communication.

Chapter 6

Sedimentation

In Chapter 5, mixing and coagulation were identified as processes that are operated as barriers against the passage of particulate contamination. Proper operation of these barriers was explained with regard to design considerations: An operator uses the right amount and types of chemicals for the proper amount of time, and with sufficient mixing energy, and expects good results.

In this chapter, sedimentation as a unit process barrier is examined. Here, too, the process has specific design and operating parameters that, if followed, will help the operator produce optimized results. In addition, *performance goals* related to sedimentation will allow the operator to determine the effectiveness of the operation. Performance goals are water quality-oriented results that operators expect to achieve most of the time under all operating conditions. These goals have common distinctions. Any operator on shift easily determines these values at any time; they are realistic for the plant operation—i.e., they are reasonably achievable, they are written and published for all to see, and they possess the "buy-in" of the entire staff and administration. In fact, the goal-setting process is a means to an end. Plants that set goals seem to outperform those that do not.

The performance goal normally associated with the sedimentation process is production of a sedimentation-basin effluent with low turbidity. Particle counts (by bin size classification) are also used to set goals for sedimentation. Because the purpose of a sedimentation basin, or clarifier, is to reduce the incoming turbidity to a fraction of its original value, goals are normally set based on the occurrence and magnitude of source water turbidity.

It is important to note that to optimize sedimentation, goals need to be established for each sedimentation basin rather than relying on goals for a composite of all the basins. Often when there are several sedimentation basins, one may have problems, but the composite results make detection difficult.

An examination of the performance goals that have been set by the Partnership for Safe Water illustrates this point. The Partnership goals for sedimentation are

- settled water turbidity with less than 1 ntu 95% of the time when source water turbidity is less than or equal to 10 ntu and
- settled water turbidity less than 2 ntu 95% of the time when source water turbidity is greater than 10 ntu.

Other sedimentation performance goals may be important to operators. It is not uncommon for operators to set sedimentation performance goals (reductions) for iron (Fe) and manganese, oxidant demand, total organic carbon (TOC), heterotrophic plate count, and color. Precipitative softening plants will have performance goals that relate to hardness and alkalinity and perhaps torque readings or settleability values for sludge blankets in clarifiers (see Chapter 9).

These goals suggest an important operational philosophy: Regardless of the incoming source water contaminant, the contaminant output from the sedimentation basin should not deviate significantly and should always be minimal, especially for particulates. There should be very little correlation between source turbidity and settled water turbidity if the basins are designed properly.

PROCESS DESIGN

Sedimentation (clarification) depends on gravity to remove, or separate, solids from water. Basins constructed for this purpose can be rectangular, square, or circular. In a conventional water treatment plant (WTP), sedimentation is situated after flocculation and before filtration. The basins are designed to allow water to flow through them very slowly, and care is taken to ensure minimum turbulence at the inlet and outlet.

Various manufacturers' designs are available to minimize short-circuiting through the basin; however, operators should expect these designs to be only partially effective. The sludge, or residuals, that accumulates on the bottom of the basin must be periodically or continuously removed to waste. The outlet of the basin is usually equipped with weirs, or launders, that collect the clear water from the top of the basin and carry it to the filters. Outdoor basins must be covered; otherwise, they will be subject to the elements. Uncovered basins are known to undergo the phenomenon akin to lake turnover when the sun shines on part of the basin and heats the water enough to cause a difference in viscosity. This common problem is seen often on circular basins at high rates on sunny days.

Detention Time

Simple clarifiers have basic design parameters that operators should understand. They are built with a minimum flow-through time or detention time. This detention time is the theoretical amount of time that a quantity of water would take to pass through the basin and exit the outlet. Most clarifiers are constructed to have 4 hours of detention time at the maximum flow rate of the plant, although groundwater-softening plants may have basins that are designed with less detention time. Detention time is calculated using the following formula:

$$\text{Detention Time} = V / Q \qquad \text{(Eq 6-1)}$$

where Q is the flow rate, in cubic feet per minute, and V is the volume of the tank, in cubic feet.

To learn theoretical detention time calculations, operators can envision the amount of time that it takes to fill or empty a tank at a given flow rate. For example, at a flow of 5 mgd (7.75 ft^3/s), how much time does it take to fill a rectangular tank to a depth of 15 ft (6 m) if that tank is 180 ft (30.5 m) long and 43 ft (17.7 m) wide, starting from empty? Using Eq 6-1, convert the flow rate from cubic feet per second to cubic feet per minute. This is done by multiplying 7.75 ft^3/s (3,480 gpm) by 60, which yields a flow rate of 465 ft^3/min (15.3 L/s). The volume of the tank is computed by multiplying the length, width, and depth, which yields a volume of 116,100 ft^3. Finally, divide the volume by the flow (116,100/465) to obtain a detention time of about 249.6 min, which is a little more than 4 hours.

This example calculation can be misleading. As discussed in the previous chapter, tanks such as clarifiers and sedimentation basins exhibit considerable short-circuiting; i.e., some of the water entering the tank escapes to the other end in a time much shorter than the theoretical detention time. Detention time is an *average* time for the water to flow through the tank. Operators need to understand that changes made to the water entering the tank in the example described here will not necessarily take 4 hours to exit the other end. This is important to understand. A mistake made in coagulation, for example, is likely to affect the filters in a time much shorter than 4 hours.

Surface Overflow Rate

Another useful design parameter for clarifiers is the surface overflow rate (SOR), which is given in gallons per minute per square foot, or centimeters per minute.

Simple basins are usually designed for an SOR of 0.5 $gal/min/ft^2$ (1.2 m/h). Basins that are operated in excess of the design SOR tend to load the filters with greater amounts of suspended solids, which can reduce filter run times and increase the probability of turbidity breakthrough in filters. The formula used to calculate the SOR is

$$\text{SOR} = Q/\text{Square Feet of Surface} \qquad \text{(Eq 6-2)}$$

where Q is the flow rate, in gallons per minute, and square feet of surface is the area of the basin. Note that the depth of water in the clarifier is not used in this calculation. This design parameter works for circular as well as rectangular basins. In the previous example, the SOR would be calculated by multiplying 180 ft by 43 ft (30.5 m by 17.7 m), which is 7,740 ft^2. The flow rate, Q, is 5 mgd, or 3,472 gpm (464 ft^3/min). The SOR then is 0.45 gpm/ft^2

(3,472/7,740) (1.47 m/h). Properly designed sedimentation-basin launderers help to control SOR.

Note that SOR can be expressed as the traditional flow rate per square foot, or as a velocity or speed that water is moving through a basin or channel. The former is used by operators to mathematical calculations of detention time, while the latter is very useful while doing jar tests. Here is an example:

> An operator needs to calculate the overall SOR of 2 basins in service when the flow rate through the plant is 4,164 gpm (6 mgd, or 558 cubic feet per minute). Each basin is 37.75 feet wide by 120 feet long by 14.8 feet depth. What is the SOR in gpm per square foot?
>
> *Answer*: 37.75 feet × 120 feet is approximately 4,530 square feet. So two basins would provide twice the area, or about 9.060 square feet. The overflow rate can then be calculated as 4,164 gpm divided by 9,060 square feet = 0.46 gpm/square foot.

What velocity is this in centimeters per minute?
$0.46 \times (1\ ft^3/7.48\ gal) \times (30.48\ cm/ft) = 1.87$ cm per minute
$0.46 \times 4.07 = 1.87$ cm per minute

In WTPs with multiple basins, operators should carefully manage the process so that these two design parameters are not exceeded. All sedimentation basins must be taken out of service periodically for cleaning and maintenance, even if the units are provided with automatic sludge-removal equipment. When basins are removed from service, operators should calculate the resulting detention times and overflow rates of the remaining basins and adjust the operation as necessary. A basin outage should be a *planned event*, and all operations staff should be aware of its ramifications.

Weir Overflow Rate

Weir overflow rates (WORs; not to be confused with SOR) are designed to not exceed 20,000 gpd/ft (250 $m^3/d/m$). Weirs at the end of the rectangular basin are designed to allow the top few inches of clarified water to be skimmed and conduct it to the filters. They are often double-sided and built with V-notches and can be positioned perpendicular to or in line with the basin flow direction. An overflow pipe is provided so that a maximum water level is maintained and not exceeded; this provides a maximum water level to the filters.

Because the weirs must be placed on top of the water, they subtract some of the total surface area of the basin and therefore cause the water to speed up as it approaches the weirs and flows into it. The last-minute increase in approach speed tends to have a pulling effect. If that tendency to pull water gets too strong, it can literally pull previously settled floc from the bottom portions of the basin and carry it over to the filters. As mentioned, to counter this effect, design engineers will provide enough weir length so that the WOR doesn't exceed about 15 gal/min/ft.

The formula used for WOR is:

$$\text{WOR} = Q/\text{Length of Weirs, ft}$$

Flow-Through Velocity

Whereas SOR uses the length and width of the basin to calculate surface area of a rectangular unit, the flow-through velocity uses the depth and width of the basin. This calculation speaks to the dramatic reduction in velocity of the water, which needs to be slow enough to allow the flocs to settle over time. For a standard rectangular basin, the design engineer will try to keep the maximum velocity at or near 0.5 ft/min. This simple calculation is:

$$\text{Velocity} = \text{Flow Rate, cfm/ft}^2 \text{ of Basin Inlet Area}$$

Using the dimensions in the previous example, we can calculate the flow-through velocity as:

$$\text{Velocity} = 464 \text{ ft}^3\text{/min/(43 ft} \times \text{15 ft)} = 0.71 \text{ ft/min}$$

This is a bit aggressive. The operator of this basin must be especially aware of the challenges of taking a basin out of service for maintenance during periods of high plant flow rates.

INLET AND OUTLET ARRANGEMENTS

Inlet devices must distribute the water equally and with uniform velocity. This is accomplished effectively if a baffle wall is placed just after the inlet. Outlet weirs or troughs should be checked for levelness periodically. An unlevel weir will pull water too aggressively from one side or section than from other parts of the weir.

TYPES OF BASINS

Conventional sedimentation basins have a rectangular-feed or center-feed configuration (Figure 6-1). Rectangular basins may be constructed of concrete or steel and are designed so that flow velocity is parallel to the basin length. The velocity through the basin should not exceed 0.5 ft/min (0.15 m/min). To achieve even flow distribution across the basin, inlets are provided that, when kept clean and free of impediments, will minimize the eddies and currents that cause short-circuiting. Outlet weirs that can also help to minimize short-circuiting are provided. The purpose of the weir is to collect clear water from the top portions of the basin for transport to the filters. An overflow pipe is provided so that a maximum water level is maintained and not exceeded; this provides a maximum water level to the filters. Basins will have various sludge collection and removal systems (see discussion later in the chapter).

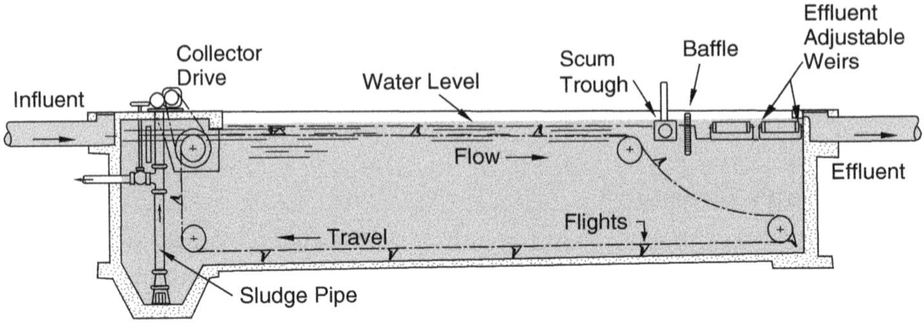

Source: AWWA 2010

Figure 6-1 Typical rectangular sedimentation basin

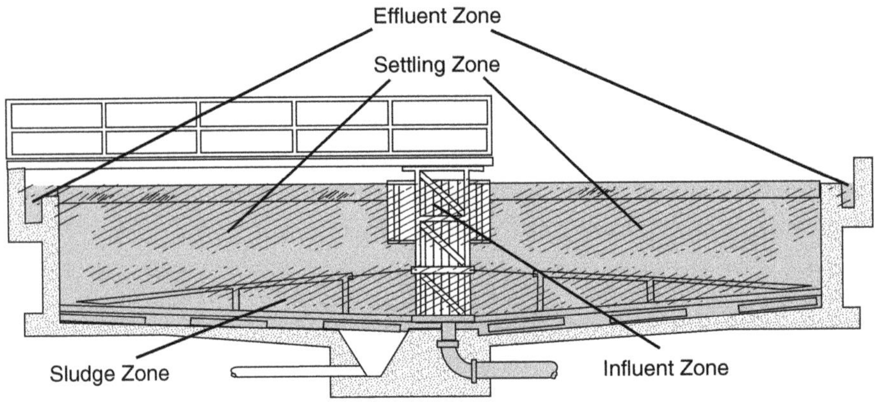

Source: AWWA 2010

Figure 6-2 Zones in a sedimentation basin

Conventional center-feed basins can be round or square. The settled water flows from the center to the outside (radial flow), and, like rectangular basins, these basins are designed to keep an even flow distribution. Basin bottoms are conical, which helps the sludge-collection process. Center-feed basins can have peripheral or spiral flow.

Sedimentation basins have four zones, each with a distinct function (Figure 6-2). The *influent zone* decreases the velocity of the incoming water and distributes it evenly. The *settling zone* provides a calm area, which is necessary for suspended particles to settle. The *effluent zone* is the area where the transition from settling to effluent flow takes place. Eddies and currents that could stir up sediment from the bottom of the basin must be minimized in this zone. Finally, the *sludge zone* is the area where sludge particles are accepted and kept separate from other particles in the settling zone. These zones are not well defined; it is difficult to determine where one zone stops and the other starts. The zones are fluid and can shift from area to area. It is important to know that the settling zone is always affected by the actions in the other three zones, and the quality of the water leaving the basin will be determined by those actions.

High-Rate Processes

Several variations of the clarification process have been developed to allow the operator to produce water at higher rates than are possible with conventional processes. Units that can operate at higher SORs accomplish the task in a satisfactory fashion and may take up less space in the plant. Most processes provide shorter response times for operators to correct treatment problems, but overall, maintenance efforts and demands are comparable to those of conventional processes.

Solids-Contact Units

A clarification unit that combines coagulation, flocculation, and sedimentation in a single basin is called a *solids-contact unit* (SCU) (Figure 6-3). Typically, these units rely on the recirculation of sludge into the process and are usually operated to maintain a sludge blanket. These units are used in the lime–soda process because the chemistry of softening is enhanced

Source: Aqua Ohio—Massillon Division

Figure 6-3 Solids-contact units at the Massillon, Ohio, WTP

by contact with previously precipitated calcium carbonate ($CaCO_3$). An SCU must be properly operated to maintain a good sludge blanket. This is accomplished by systematically withdrawing a portion of the sludge. Automatic timer controls are usually provided, and the operator must provide the necessary settings. These units, designed to operate at SORs of 1.0 gpm/ft^2 (2.4 m/h), may be designed for even higher rates if the unit is used for lime softening, but they do not handle magnesium (Mg) hardness as well as they do calcium (Ca) hardness. Calcium carbonate settles more quickly than magnesium hydroxide. The main drawbacks of these units are sensitivity to increases in flow rates and to temperature variations. Both conditions can upset the sludge blanket and result in floc carryover.

Tube and Plate Settlers

Particles suspended in water must settle to the bottom of a basin for the sedimentation process to be effective. In conventional settling basins, this often means that a particle must travel 15 ft (4.5 m) or more before it is truly removed from the process. As gravity pulls the particle to the bottom of the basin, it resists settling because of the horizontal flow of water and so may be carried to the filters. Consequently, basins must be large enough to provide long detention times.

One way to reduce the distance that a particle must travel is to insert tubes or plates into the basin at angles; a practice that *increases the effective surface area of the unit.* The result is shallow-depth sedimentation, which can increase the effective SOR of the basin two to four times. Typical SOR rates for basins equipped with tubes or plates are 2.0 gpm/ft^2 (4.8 m/h). Tubes or plates are designed as either horizontal or as steeply inclined, depending on the application. See Figures 6-4 and 6-5.

This technology is frequently used in small package plants. Horizontal tube settlers are actually slightly inclined (perhaps 5°) and work dependently with the downstream filter. Sludge accumulates in the tubes until the filter is backwashed. When the filter is drained, the water level in the settler is lowered, and the sludge is carried to waste.

RETROFIT OF STANDARD BASINS

Steeply inclined tubes or plates are designed into basins with varying degrees of incline. When they are inclined at less than 45°, the method of sedimentation is generally capture and storage. For this reason, these units must be cleaned periodically to prevent overaccumulation of solids. In some plants, tube settlers have accumulated enough solids to collapse under the weight.

Cleaning is usually accomplished by slowly lowering the water level in the basin, allowing the solids to slough to the bottom, where they are wasted. At an incline greater than 45°, solids do not accumulate on the tubes but rather move down the tubes in a continuous fashion. Tubes and plates are built and shipped as modules of varying size and therefore lend

Source: Lake County, Ohio, Department of Utilities

Figure 6-4 Tube settlers at Lake County East WTP

themselves to retrofitting into existing basins. To gain more capacity, many WTPs have resorted to this type of retrofit rather than building new and expensive sedimentation basins.

These arrangements rely on an upflow direction of water; i.e., the water to be settled is introduced to the bottom of the plates or tube settlers and allowed to flow up through them. If operated properly, the water velocity is low enough to allow the solid particles to settle. Another arrangement is to install parallel plates into the basin with the water to be settled directed downward to the bottom of the basin and then upward into the plates. These plates are placed at an angle and are referred to as *inclined plate settlers,* and are usually inclined at 55°.

Superpulsators

A superpulsator is a modification of a sludge blanket clarifier coupled with lamella plates, which enhances coagulation. SORs as high as 5 gpm/ft^2

Source: Aqua Ohio–Ashtabula Division

Figure 6-5 Plate settlers at the Ashtabula, Ohio, WTP

(12 m/h) have been reported for these units. The water entering the clarifier is "pulsed" by the application of a vacuum created in an upstream vacuum chamber. The vacuum pulls the water into the chamber, and, when the water rises to a certain level, the vacuum is released. The water then surges into the clarifier through the distribution conduit and laterals, causing the sludge blanket to expand uniformly. The chemically treated water must pass through the sludge blanket, where particle contact takes place. With each surge, some of the residuals are allowed to overflow into a sludge concentrator zone, where they are thickened and sent to waste. The supernatant flows up through the lamella plates, which have been modified with deflectors. The surges create small eddy currents that are redirected by the deflectors into the flocculated sludge. The heavier particles in this zone settle out, and the clarified water is collected and sent to filtration.

Sand Ballasted Clarification

This process includes distinct rapid mix, maturation, and clarification zones. The technology uses select silica sand added to the water as "seed" particles to enhance agglomeration and the settling characteristics of the particles formed. A metal salt coagulant is added to the rapid mix zone, which provides intense mixing energy for 2 min or less. The maturation zone is an upflow basin or the draft tube and is where the polymer-coated sand is introduced to the process stream. This stream follows an underflow passage into the clarification zone. This deep settling zone, with a detention time of about 6 min, is where clarified water is collected over the lamella or tube settlers. Solids recirculation pumps return the solids to the hydrocyclones. These units provide shearing forces that separate the sand from the treated solids and return the significant portion of the sand to the process. The remainder of the sand and residuals is wasted.

Dissolved Air Flotation

Dissolved air flotation (DAF) is a high-rate process that has been successfully applied in many parts of the world for decades. Like most high-rate processes, it has more system components compared with a conventional facility; however, it is reliable and robust and does not necessarily require more operator sophistication. In fact, numerous utilities have recently installed DAF systems to replace conventional plants and have made the transition rapidly and more simply than anticipated. DAF systems typically operate at hydraulic overflow rates in the range of 6–10 gpm/ft^2 (14.7–24.5 m/h).

DAF involves rapid mixing and flocculation followed by clarification; sludge is brought to the top of the basin, rather than to the bottom, as with conventional treatment. This is accomplished by recycling a portion of the clarified water (about 10%), saturating it with air under pressure (roughly 70 psi), and delivering it to the base of the flotation tank, where a pressure drop occurs through a series of nozzles or valves. Because of the pressure drop, air is released from solution and mixed with the incoming flocculated water. The bubbles attach to the floc, which then rise to the top and rear of the tank over an inclined baffle. A sludge, or "float," layer develops at the top of the flotation tank, which is periodically removed by mechanical (skimmer) or hydraulic means. The objective of DAF is to produce small, pin-sized floc that is readily amenable to flotation.

DAF was originally used to treat cold, low-alkalinity waters that were prone to algal blooms. DAF has also been successfully used to treat water with more diverse qualities, especially recycled backwash water, which tends to be highly turbid and variable.

In a pilot-scale study performed at the Cleveland, Ohio, Division of Water, Morgan WTP, DAF was shown to treat spent filter backwash water (SFBW) at a hydraulic overflow rate of 6 gpm/ft^2 (14.7 m/h). This was eight times the rate used by conventional clarification with polymer addition in the parallel study and comparable to rates used in potable water treatment. The average untreated turbidity of the SFBW was 25 ntu and ranged between 10 and 60 ntu during the study. In the study, conventional clarification with polymer (at an overflow rate of 0.8 gpm/ft^2 [2 m/h]) reduced the clarified effluent turbidity to 2 ntu, while the DAF system reduced it to less than 1 ntu on average.

Source: Nick Pizzi

Heavy floc carryover can reduce filter run length.

Figure 6-6 Floc carryover and ice formation on SCU

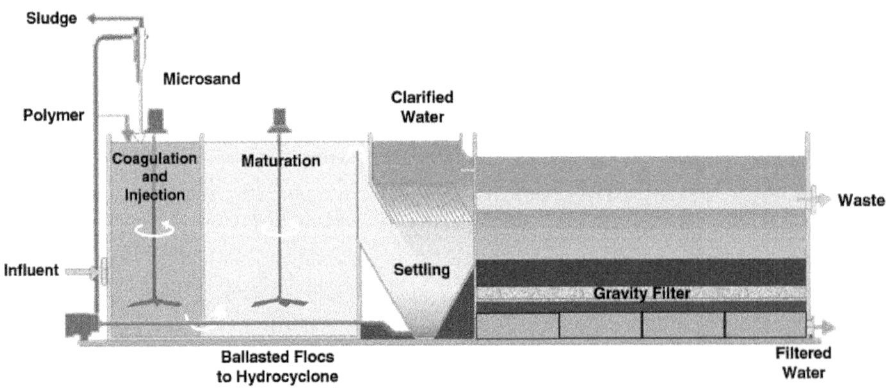

Figure 6-7 Vacuum sludge-removal installation in sedimentation basin at Lake County East WTP

Because of its ability to remove low-density particles, DAF has achieved a 2-log removal of *Cryptosporidium* oocysts, which compares to a removal rate of less than 0.5 log using conventional sedimentation.

OPERATIONS AND MAINTENANCE

Sedimentation is an important barrier against particulate passage into the filters. When designed and operated properly, clarifiers remove a tremendous amount of particulates in a relatively short period of time. It is not uncommon for sedimentation basins to remove 90% or more of the source water turbidity. Conventional basins tend to be very expensive to design and construct. As a result, in larger WTPs, conventional basins barely meet minimum design standards for detention time. When taken out of service, they represent a large percentage of capacity loss. In a small plant with only two sedimentation basins, 50% of the capacity is lost each time a basin is down for service. Explained another way, 50% of the process that removes 90% of the solids is temporarily lost in that instance.

As previously mentioned, sedimentation basins should be cleaned manually, preferably twice each year. Most plants schedule this maintenance just before and just after the summer high-pumping months. Plant operators in northern climates prefer not to remove aboveground basins in winter because colder water makes settling more difficult. Also, the chilled basin walls make it very uncomfortable for operators to work, even when the basins are covered.

Manual cleaning is important because harmful and nuisance bacteria and organisms can collect in the slimes that adhere to the basin walls and appurtenances. These slimes, or biofilms, accumulate on rough surfaces and in crevices where chlorine (Cl) has difficulty reaching. Particularly vulnerable are the interfaces between the water surface and the air surrounding it. Ever-present scum lines, or "rings," make an excellent shield for organisms such as Legionnaires' bacteria. Operators need to blast these areas with water from high-pressure hoses to release the bacteria. Operators must wear paper masks to prevent inhalation of the mists generated during blasting.

To reduce chlorinated byproducts, many plant operators have reduced or eliminated prechlorination. At these plants, it may be wise to periodically chlorinate the basin for a few hours. This maintenance procedure, referred to as "shock" by operators, is typically performed in uncovered basins that are exposed to direct sunlight. These basins tend to accumulate algae that can cause taste-and-odor problems. Another way to rid a basin of slime growth is to coat the walls with a mixture of copper sulfate and lime, usually at a concentration of 10 g/L. The basin is drained and the mixture applied to the walls with a brush.

Sludge often accumulates on the basin floor, even when automatic sludge-removal devices are used. Most basins have areas where cleaning equipment cannot reach. Sludge can age in these spots and "go septic," i.e., become devoid of oxygen (O) and produce foul-smelling gases. This is to be avoided.

It is a good idea to perform physical inspection of the basins while they are emptied. Basins with notched launderers should be checked to see that the troughs have not become misaligned because of freezing conditions. Unleveled troughs will create hot spots of localized high overflow rates and draw too many solids from the depths of the basin. This can cause filtration problems. If catwalks or walkways are a part of the basin, operators should check for rust or paint chips or grease that can drip and contaminate the water. This is also a good time to maintain any moving parts.

Note the baffle wall that was installed during the retrofit, with tube settlers at the effluent end of the basin. This wall forces all water to the bottom and then up through the tubes, where solids are trapped.

Safety

Safety measures are important when cleaning sedimentation basins. Most interpretations of Occupational Safety and Health Administration rules classify sedimentation basins as confined spaces. Consequently, these rules would apply to cleaning procedures of or for any entry into the unit. The use of high-pressure water hoses can present a hazard to the operator. Wet, slippery floors and the loss of control of a hose are dangerous to the operator and any other occupants. Also, because unexpected flow of water into the basins from the inlet source is a form of energy release, many operators follow lockout–tagout procedures before entry. An operator working in a deep basin with his or her back to the wall may not be able to hear the rush of water that results from an inadvertent valve opening until it is too late. Many plants keep a life ring or pole handy in case a worker falls into the basin. Moving parts, especially in a chain-and-flight system, present a hazard to workers. Such machinery should have guards placed on them. Guardrails that surround the basins are required. Safety practices are explained further in Chapter 14.

Residuals

All basins, whether conventional or high rate, require sludge disposal. This is usually the most troublesome operation in a WTP. In the past, operators commonly discharged residuals to a stream or lake. However, discharge regulations, both for the environment and for WTPs, now restrict an operator's ability to waste residuals with impunity. The process of sludge disposal is now as tightly controlled as any other process in the plant. Residuals are removed manually or with an automatically controlled vacuum or raking systems.

Residuals are composed of all the unwanted material that has been removed from the water and all of the compounds that are created by addition of treatment chemicals (floc, lime solids, etc.). Because this chapter is about sedimentation, residuals from brines used in ion exchange or wastes from membrane applications are discussed in other chapters.

Alum sludge is the most common residual produced in WTP sedimentation basins, but ferric sludge is also common. Both wastes come from the sedimentation basins at low concentrations by weight: normally less than 2%. It is inefficient to dispose of this low concentration of sludge, and so WTP operators dewater and concentrate the residual.

Dewatering is the process of removing water from the thin sludge, allowing the solid material to remain (thickening) for ultimate disposal. It is difficult to dewater alum and ferric sludge because the water is chemically bound to the aluminum and ferric hydroxide floc. These flocs are gelatinous and thus difficult to handle. As a rule of thumb, the thicker the sludge, the less space it takes up, and so less storage space will be needed. Additionally, the thicker the sludge, the less expensive it is to dispose. WTPs and landfills usually charge by volume of material they receive (there may be other surcharges for strength of solids or for special considerations because of the nature of the sludge). When disposing of thin sludge, therefore, the WTP operator is also paying to dispose of a larger proportion of water than may be necessary.

Sludge drawn from sedimentation basins varies with the coagulant used, of course, but also with the nature of the material found in the source water. Sludge produced in plants that draw water from lakes and other sources with relatively low turbidity is very gelatinous, whereas sludge produced from a source high in silt and organic content is not as gelatinous and thus more easily handled. The percent solids concentration of the sludge also affects its characteristics—low-solids sludge is more liquid than high-solids sludge, which tends to handle like clay.

In many WTPs, sludge from sedimentation basins is pumped to a gravity thickener that is designed to dewater the waste to a specific concentration before disposal. This gravity system allows the solids to settle to the bottom and provides for removal of the clear supernatant at the top of the unit. Supernatant is either returned to the process or sent to waste. The thickener also serves as an equalization tank, allowing for continuous or intermittent wasting of solids that might otherwise burden the receiving sewers or treatment facility.

Further treatment of sludge after thickening can take place in the plant or at the disposal site. WTPs often treat thickened sludge with polymers or lime and may subject the sludge to flotation or belt-press thickening to obtain the desired concentration. When sand drying beds, lagoons, or freeze–thawing beds are used, the process is referred to as *nonmechanical dewatering*. Mechanical dewatering is usually accomplished with vacuum filtration, belt presses, or centrifuges. All of these processes are effective but costly. Each WTP needs to determine the most cost-effective procedure for its situation, taking into account all economic factors at its disposal. Local regulations will also be a factor.

For simplicity and operator convenience, the sand drying bed method of solids handling is difficult to beat. Costs of the process vary by site and size, but beds with greenhouses built around them to trap heat, and with fans to hasten drying, show great promise for freeing plant operations from the burgeoning costs imposed on them for residuals disposal when discharging to wastewater systems.

Some utilities with warm and sunny climates have been using simple sand beds for years. Ongoing research is leading engineers to design these systems for utilities in varying climates. One process refinement is simple in its application: Thickened sludge is pumped to the bed as polymer is injected into it. Sludge is loaded onto the beds at rates of about 3 dry lb of solids per square foot of bed. Polymer dosages are typically found to be 3 lb of polymer per dry ton of sludge. The cakes that are formed can quickly dry to 30% or more in a few days and are easily handled and disposed.

Sludge Calculations

If the concentration of solids coming from the sedimentation basins is known, the amount of solids that the thickener will produce can be calculated using the formula:

$$V_2/V_1 = P_1/P_2 \qquad \text{(Eq 6-3)}$$

where V_1 and V_2 are the beginning and ending volumes of the sludge and P_1 and P_2 are the beginning and ending concentrations of sludge, expressed as percent. This formula assumes that the specific gravity of the sludge varies little in typical dewatering processes. As an example, suppose that 100,000 gal of sludge is removed from a sedimentation basin at 0.2% (2,000 mg/L) solids and is sent to a thickener where it will be concentrated to 5% (50,000 mg/L) before being transferred to a tank truck for hauling. If the tank truck has a 4,000-gal capacity, how many trips will be necessary?

Answer: $V_2 = P_1 \times V_1/P_2 = 0.2 \times 100{,}000 \text{ gal}/5 = 4{,}000 \text{ gal}$

One truckload will be sufficient.

Operators use other formulas to predict the amount of sludge that will be produced by the chemical coagulation and/or precipitation processes. These formulas are for alum and ferric salts and for softening chemicals such as lime and soda ash.

$$\text{For alum, } S = (8.34Q)(0.44\text{Al} + SS + A) \qquad \text{(Eq 6-4)}$$

where:

S = sludge produced, in pounds per day
Q = plant flow, in million gallons per day
Al = dry alum dose, in milligrams per liter as 17.1% aluminum oxide (Al_2O_3)
SS = source water suspended solids, in milligrams per liter

A = solids from other treatment chemicals, such as powdered activated carbon, etc., in milligrams per liter

$$\text{For iron salts, } S = (8.34)(2.9\text{Fe} + SS + A) \quad \text{(Eq 6-5)}$$

where Fe is the iron dose, in milligrams per liter as Fe.

$$\text{And for softening, } S = (8.34\ Q)(2.0\text{Ca} + 2.6\text{Mg}) \quad \text{(Eq 6-6)}$$

where Ca and Mg are the calcium and magnesium hardness removed, each expressed as milligrams of calcium carbonate per liter.

Example Calculation

The *SS* portion (source water suspended solids) of the preceding formulas is either measured in the laboratory or approximated using a "*B* value." The *B* value is a ratio of the suspended solids in the source water to the turbidity of the source water. It is usually found to be about 0.7–2.2 but can be much higher if large amounts of TOC are present. An operator's ability to quickly analyze for turbidity makes the *B* value a great help in predicting solids production. Each WTP needs to develop a history of solids measurement to rely on a usable *B* value.

Using Eq 6-4, calculate the pounds of solids produced per day if a treatment plant processes 25 mil gal of water using 15-mg/L dry-basis alum. The source water turbidity is 44 ntu, and the historic *B* value for this plant is 1.5. The plant is adding 5-mg/L activated carbon for taste-and-odor control.

Multiply the *B* value by the *SS*: 1.5 × 44 = 66. The calculation becomes

$$(8.34 \times 25)((0.44 \times 15) + 66 + 5) = (208.5) \times (77.6) = 16{,}180 \text{ lb}$$

TURBIDITY TESTING

As previously mentioned, the performance of a sedimentation basin is commonly measured using turbidity output. It is a good practice to determine the turbidity from each basin rather than from the combined output of a series or bank of basins. A poorly performing unit can be masked by the dilution of good water from the other basins in the bank. Some operators prefer to measure the incoming and outgoing turbidity or particle count of each basin. The difference in the two readings is measured against a preset goal, either as a percent removal or as a log removal. When the goals are not met, remedial action is indicated.

BIBLIOGRAPHY

AWWA. 2011 (6th ed.). *Water Quality and Treatment* (J.K. Edzwald, editor). New York: McGraw-Hill.

AWWA. 2010 (4th ed.). *Principles and Practices of Water Supply Operations—Water Treatment*. Denver: AWWA.

Cornwell, D.A., M.M. Bishop, R.G. Gould, and C. Vandermeyden. 1987. *Handbook of Practice: Water Treatment Plant Waste Management.* Denver: AWWA and WRF (Water Research Foundation).

Cornwell, D.A., M.J. MacPhee, N.E. McTigue, H. Arora, M. LeChevallier, and J.S. Taylor. 2001. *Treatment Options for Removal of* Giardia, Cryptosporidium, *and Other Contaminants in Recycled Spent Filter Backwash Water.* Denver: AWWA and WRF.

Cornwell, D.A. and D.K. Roth. 2008. *Altering Environmental Conditions to Enhance Non-Mechanical Dewatering of Water Treatment Plant Residuals.* Denver: WRF.

EE&T Inc. (Environmental Engineering & Technology Inc.). 2000. "Preliminary Evaluation of Morgan WTP Residuals Handling Facilities." Newport News, Va.: EE&T Inc.

Edzwald, J.K., J.E. Tobiason, H. Dunn, G. Kaminski, and P. Galant, 2001. "Removal and Fate of *Cryptosporidium* in Dissolved Air Drinking Water Treatment Plants." *Water Science & Technology*, 43(8):51–57.

HDR Engineering Inc. 2001 (2nd ed.). *Handbook of Public Water Systems.* New York: Wiley.

Linder, K. and B. Martin. 2015. *Self-Assessment for Water Treatment Plant Optimization.* Denver: AWWA.

MacPhee, M.J. 2001. Personal communication.

MacPhee, M.J. and D. Cornwell. 2000. "Critical Assessment of Alternatives for Treatment of Spent Filter Backwash Water." In *Proc. Water Quality Technology Conference.* Denver: AWWA.

Pizzi, N.G. 2010. *Water Treatment Plant Residuals (Pocket Field Guide).* Denver: AWWA.

Pizzi, N.G. and M.L. Rodgers. 1998. "Using Particle Count Data to Improve Operations." In *Proc. Water Quality Technology Conference.* Denver: AWWA.

Water Supply Committee of the Great Lakes–Upper Mississippi River Board of State and Provincial Public Health and Environmental Managers. 2007. *Recommended Standards for Water Works.* Albany, N.Y.: Health Research Inc., Health Education Services.

Chapter 7

Filtration

Filtration is the fourth unit process that makes up conventional treatment for particle removal; the others include coagulation, flocculation, and sedimentation. Filtration is used in water treatment plants (WTPs) to remove particulate material from the water and store it for eventual disposal. Particulates include those already present in the source water, such as clay and silt particles, bacteria, protozoans, viruses, and organic substances, as well as those generated during the treatment process, e.g., when alum, ferric salts, calcium carbonate ($CaCO_3$), and magnesium hydroxide solids are used in the softening process. In some plants, appreciable amounts of iron (Fe) and manganese may also be present.

Surface WTPs (and those considered to be under the influence of surface water) are required to filter water to acceptable levels of turbidity (see Chapter 1). AWWA's *Water Quality and Treatment* (2011, page 10.4) speaks to the importance of this unit process:

> It can be argued that particle filtration processes are possibly the most critical component in potable water treatment facilities because of their ability to remove microorganisms, especially pathogens that cause acute health concerns.

The quote reminds operators that their operational skills related to filters are a matter of public health. It is not difficult to understand why so much time and effort have gone into the study and design of the filtration process and why there are many different types of filtration processes found in WTPs. Most common are slow sand filtration and rapid sand filtration units, which are granular bed processes. Diatomaceous earth (DE) beds, which use a permeable membrane, are also used. Membrane processes are also common and are covered in Chapter 10.

GRANULAR MEDIA FILTER THEORY

Any remaining suspended solids from the clarification process need to be captured by filtering the water. To accomplish this, the typical WTP will use filters constructed of fine grains of a durable material (media) that can

Source: Cleveland, Ohio, Division of Water

Figure 7-1 Filter gallery at Cleveland's Baldwin WTP

be layered into a box or vessel (Figure 7-1). The box will be equipped with some type of support system that can retain the media while allowing the water to pass through. Filters consume energy, and enough of it must be supplied to overcome the drag forces that the water experiences as it passes through. This force needed is called "head," and it can be supplied by the WTP pumping system or well pumps or supplied by gravity. If the energy comes directly from a pump and the vessel is closed, we refer to the unit as a "pressure" filter. When the vessel is open to the atmosphere, we refer to it as a "gravity" filter. Gravity filters can be classified by type as either slow sand filters, rapid sand filters, or high-rate filters.

Most granular media filters require "pretreatment"—coagulation and other conditioning of the source water before being applied to the unit. The exception is the slow sand filter. This pretreatment—destabilization and coagulation and settling—was discussed in Chapter 5.

Slow Sand Filters

Slow sand filters were the first type of gravity filter used for water treatment (Figure 7-2). This type of filter, which is always open to the atmosphere, operates at very low filtration rates and does not accept water that has been treated with a coagulant. These filters, which use small-sized sand, compared with rapid sand units, operate at a low filtration rate (0.05 gpm/ft^2 is typical). Gravity allows the water to flow downward through the bed. This results in solids being removed and stored almost entirely in the top few inches of the sand bed. This layer of sand becomes laden with particles and dead and living microorganisms and serves as the dominant filter media as the filter cycle progresses. Filter run lengths for these types of filters are

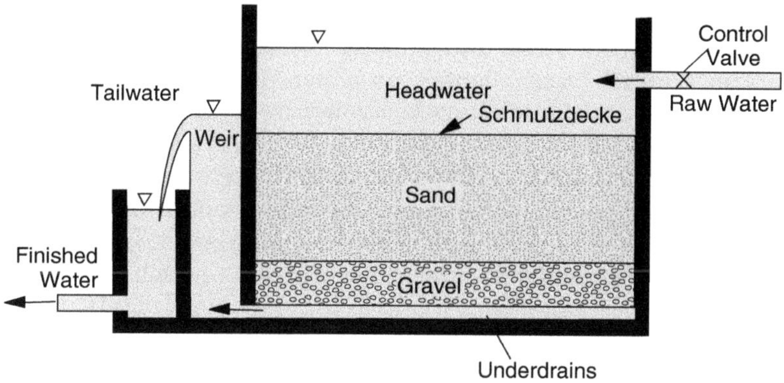

Source: AWWA 2011

Figure 7-2 Slow sand filter

usually one to six months. When head loss becomes excessive, the filter is cleaned by draining the bed below the sand surface and removing the top inch or two of sand.

Slow sand filters are often considered because of their simple technology—no knowledge of coagulation chemistry is required.

Slow sand filtration is most widely used by municipal WTPs (i.e., not plants that serve industry) where source water quality is stable, notably in upper New York and the New England states, as well as in the Pacific Northwest.

A slow sand filter is capable of producing a low-turbidity effluent and can effectively remove or control a wide variety of microorganisms. Studies have shown that slow sand filters removed 85% of coliforms and 98% of *Giardia* cysts in a source water at startup and performed even better as the bed matured. Although there are exceptions, slow sand filters usually produce effluent turbidities that are less than 0.5 ntu as long as source water remains stable and of good quality. Experts expect an increase in use of these units in the future, especially by small systems.

Slow sand filter design usually provides for filtration rates of 0.016–0.16 gpm/ft^2 (0.04–0.40 m/h) when sand sizes of 0.15–0.40 mm are specified. Surveys of many US treatment plants reveal that operating head losses fall within the range of 2.5–14 ft (0.76–4.3 m).

Operations and maintenance for these units include checking and recording head loss, adjusting filter production to provide for system demand, and measuring filtered water turbidities and disinfectant residuals. Cleaning, or scraping, the bed is labor-intensive, and several mechanical devices have been constructed to assist the operator in this effort. The sand that is removed is usually cleaned and stockpiled for repeated use. The bed can be scraped for several filter cycles, but eventually the bed must be replenished. This operation is called *resanding*.

Studies of the biota of biological filter have shown they include bacteria, protozoa such as rhizopods and ciliates, rotifers, cepapods, and aquatic worms. The type of biota changes as water flows deeper into the sand bed. Biota provide for the schmutzdecke, which breaks down some organic matter, and suspended particles, which may be physically strained out of the water. Water then enters the top layer of sand, where more physical straining and biological action occur. Attachment of particles onto surfaces of the sand takes place. Algae mats, and of other filter-clogging species, may form at the schmutzdecke. Some algal cells may penetrate the top inch of sand.

Among the largest organisms found in the slow sand filter bed are roundworms, flatworms, and segmented worms, which can be as large as a few millimeters in size. These worms feed on detritus and fragments of organic matter, which tend to be deeper in the bed than bacteria and protozoa, so it's important for the operator to understand that the sand bed is inhabited by living organisms. To do the job, these organisms need both food and oxygen to metabolize the food they ingest. So operators must provide a steady input of fixed food and oxygen in the filtered beds to help stabilize their population. Long-term interruptions of flow can lead to depletion of oxygen in the filter bed, causing the die-off of organisms that are key to effective slow sand filtration. Drying a slow sand filter bed is even more serious and could lead to elimination of most all the biota in the bed. Therefore, the flow of influent water to the unit should be continuous during water production. And filter bed maintenance activities need to be performed in a prompt manner to minimize downtime and harmful effects to biota. Sand that is put into the filter needs to be thoroughly washed before installation. Filters are cleaned by removal of the top inch of sand. When it is replaced, if it is not thoroughly cleaned, it can pass fine clay particles that can persist for months.

Slow sand filters bring the advantage of low cost for operations, maintenance, and initial construction, but they need large tracts of land, and freezing can be a problem. They do not require operational expertise in chemical coagulation, nor do they typically need "around the clock" oversight. They have very few moving parts that will wear down, with the exception of a mechanical scraper if present.

Rapid Sand Filters

Proper operation of rapid sand filters, also known as rapid granular bed filters, usually depends on the passage of pretreated water (see Chapter 5) through the bed at rates sufficient to drive the particulate material into the deeper layers. Sand sizes are larger than those of slow sand filters. This arrangement, coupled with the higher filtration rates, generates filter runs that are usually hours to a few days in length, as opposed to months.

Unlike the slow sand filter, which uses only the top few inches of the bed as the filtering media, the rapid sand filter may use the entire bed depth for particulate capture. Consequently, it cannot be cleaned by scraping. Cleaning, or backwashing, is accomplished by introducing water and/or air into the bottom or outlet end of the bed in sufficient quantities and with

Source: Kenneth P. Day

Figure 7-3 Filters #1 and #2: Retrofitted dual-media beds at a WTP

sufficient flow to fluidize and expand the media so that the stored particulate materials are dislodged. These particulates are then carried away in the stream for disposal or reuse.

While the rapid sand filter performed very well in WTPs across the country for many years, a lot of them have been retrofitted to dual-media higher-rate filters that can produce more unit filter run volumes (UFRVs) in the same footprint (Figure 7-3).

As mentioned earlier, rapid sand filters that are open to the atmosphere are called *gravity filters*. Those that are closed and under pressure are referred to as *pressure filters*.

Conventional Gravity Rapid Sand Filters

Rapid sand gravity filters are designed open to the atmosphere and use gravity as the driving force. An attribute of the gravity filter design is that an operator can view the filter bed and backwash actions at any time, which is an enormous advantage over pressure filters.

Rapid-rate granular media filters (rapid sand filters) can accommodate much higher filtration rates than slow sand filters because they use coarser sand (Figure 7-4). Instead of depending on the schmutzdecke for filtering

Source: Kenneth P. Day

Figure 7-4 Rapid-rate filter trough

action, the filters trap suspended matter through several inches (centimeters) or more of the depth of the filter sand. Rapid sand filters are designed so that they can be backwashed to be cleaned and restored for use. At least two filters must be provided so that one can remain in operation while the other is being backwashed. If there are only two filters, each must be capable of meeting plant design capacity.

Rapid sand filtration requires chemical pretreatment, usually with a metal salt coagulant to form floc particles that can be captured in the filters. These types of filters were the common units of the mid-20th century. They became the workhorses of the drinking water industry that served the huge and rapid increase in population after World War II. There are still many of these in service, especially found at groundwater systems that don't have to meet Surface Water Treatment Rule requirements and regulations. They're still quite common at precipitative softening plants.

High-Rate Granular Media Filters

As mentioned, fine-sand monomedia beds were eventually replaced in many installations with the dual-media design. The dual-media and trimedia designs allow for greater bed penetration by particulate material and therefore for longer filter runs at higher filtration rates. Deep-bed monomedia filters with coarse media (1-mm effective size or greater) are also being used where regulatory agencies have approved the design. Local regulatory approval of all media designs is usually required before the design is put into use. Granular activated carbon (GAC) beds are used alone or as caps on dual-media filters to remove tastes and odors and to improve organic carbon removal. They may also be used as biological filters. The use of GAC beds is discussed in more detail in Chapter 10.

Most higher-rate gravity filters achieve rates of 4–6 gpm/ft^2 (10–15 m/h) and run lengths of 16–80 hours or more. It is common for these filters to produce effluent turbidities of less than 0.1 ntu for most of the run. Operators report head losses of 2–8 ft (0.6–2.5 m) and UFRVs between 5,000 and 10,000 gal/ft^2 (205,000 and 410,000 L/m^2).

Pressure Rapid Sand Filters

Pressure filters are closed tanks or cylinders (i.e., not open to the atmosphere) that contain granular media for filtration (Figure 7-5). This allows the operator to rely on single pumping through the filter and into the system without loss of pressure, as in a gravity filter. This can result in savings in electrical costs. Also, development of negative head and the problems that accompany this phenomenon are avoided with this design. Plants with pressure filters may be easier to automate.

However, there are disadvantages of this design, including the fact that the operator cannot see the top of the bed during filtration or backwash. Also, because the tanks are cylindrical, the width of the filter media is not uniform, which can lead to uneven backwash patterns.

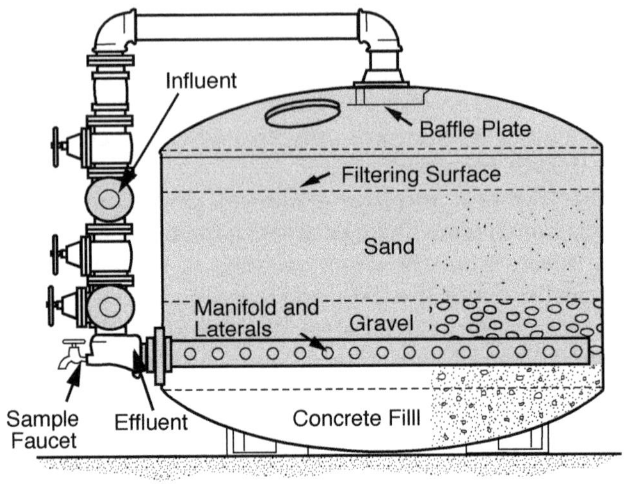

Source: AWWA 2011

Figure 7-5 Graphic of a pressure filter

The operation of pressure granular filters is similar to that for their gravity filter counterparts: Water must be pretreated, media size and depth are similar, and head loss developments are comparable. Pressure filters, which find their greatest use in industrial applications and for swimming pools, are also used sparingly for potable water production, particularly in small systems.

DE Filters

DE filters, also known as precoat filters, are an acceptable filtration technique for potable water production. DE is the skeletal remains of microscopic diatoms or algae. This material is mostly silica and inert; therefore, it is suitable to filter water. Because coagulation is not required for operation of these units, operators who possess mechanical skills are more important to the process than those who possess scientific skills. For small systems that must share staff to perform several differing tasks, this can be important.

A high efficiency for removing *Giardia* and *Cryptosporidium* has been reported at facilities that use precoat filters. Other advantages include capital cost savings (precoat filters require only a small space for installation), ease in handling waste residual, and lack of terminal turbidity breakthrough (because of surface, rather than depth, filtration). Disadvantages are that source water quality must be good and that the technology may not be applicable for removing total organic carbon.

Process operation. The process uses DE to strain particulates from water and normally does not use coagulant chemicals. A thin layer of DE coats the filter leaves, or septum, which is a plastic or wire mesh covering a hollow collection channel. The DE coating (precoat) is formed by a recirculating slurry of DE.

There are two main types of DE filter systems: pressure and vacuum. Both have advantages and disadvantages.

Untreated water is applied to the precoat filter (usually about ⅛ in. thick) with a small amount of DE added (called *bodyfeed*). Particulates in the water are removed on the filter surface, and this causes pressure to build (head loss). When the terminal head loss is reached, the filter is removed from service for cleaning. High-pressure water spray on the filter cake is used to remove the accumulated material. This waste material is discarded, and the process is repeated, starting with the filter precoat with fresh DE.

DE filtration is used for source water with a maximum turbidity of 10 ntu. The filter loading rate can be between 0.5 and 2.0 gpm/ft^2. DE filtration is very effective for removing *Giardia* and *Cryptosporidium*. Dissolved substances and color are not effectively removed by DE filtration.

GRANULAR BED DESIGNS

The most common granular filters (Table 7-1) are monomedia, dual-media, and tri-media beds that have sufficient depth to produce long filter runs. In the dual and mixed designs, anthracite or activated carbon is used along with sand that has a carefully chosen specific gravity (lower than that of reg-ular sand). This allows the larger-sized anthracite or carbon grains to settle on top of the smaller-sized sand. This large-to-small arrangement provides the longer filter runs and clearer effluent that are desired.

In dual-media filters, it is important that the size of the larger anthracite particles be compatible with the smaller-sized sand particles. When properly chosen, anthracite particles at the 90% size will be about three times larger than the sand particles at the 10% size (the effective size). This ratio, called the d_{90}/d_{10} ratio, helps to determine the amount of intermixing at the coal–sand interface. If this ratio is less than 3, a sharp interface will be created, and excessive head losses and possible mudball formation will result. If this ratio is close to 4, there is too much intermixing at the interface, and poor-quality filtrate may result. Consequently, operators should be careful when adding media or "capping off" their filters. The addition of a few inches of the wrong-sized media can defeat the effectiveness of the filter.

Table 7-1 Typical filter media design parameters

	Effective Size, ES	Uniformity Coefficient, UC
Monomedia sand	0.35–0.7 mm	1.3 to 1.8
Dual-media, anthracite / sand	0.9 to 1.4 mm / 0.35 to 0.7 mm	1.3 to 1.8 / 1.3 to 1.8
Tri-media, anthracite, sand, ilmenite	0.9 to 1.4 mm / 0.45 to 0.55 / 0.20 to 0.35 mm	1.4 to 1.75 / 1.4 to 1.65 / 1.2 to 2.0

Source: Nick Pizzi

Figure 7-6 Sieve pans used for filter media size analysis

For the best chances of doing it right, media samples should be sent to an accredited laboratory for "sieve analysis" periodically (Figure 7-6). The lab will test for effective size, uniformity coefficient, and other important design parameters for the media.

Other Filter Characteristics

Pretreated water is fed to the filter influent through a pipe or gullet that is designed to cause minimal disturbance. Free-falling or turbulent water should not be allowed to enter the filter because it may disturb the surface of the bed. Baffling may be needed at the influent to reduce any disturbance. Influent piping is designed to provide a velocity of about 2 ft/s (0.61 m/s). Effluent and wash-water pipes may be designed for velocities that are two to three times higher than velocities at the influent pipes.

A filter normally includes five valves: influent, effluent, wash-water supply, wash-water drain (sewer), and surface wash or air-scour supply (Figure 7-7). If filter-to-waste capability is added, a sixth valve may be needed. Most valves used on filters are the rubber-seated butterfly type. These valves, which are placed for accessible maintenance, are equipped with position indicators that typically are electrically controlled.

Wash-water troughs are used to carry solids away during backwash. By limiting the horizontal distance that floc particles suspended during the

Source: Lake County, Ohio, Department of Utilities

Figure 7-7 Local control filter panel at Lake County, Ohio, Bacon Road WTP

wash cycle must travel, properly spaced troughs allow for efficient removal of solids. Troughs must be placed at a level above the filter bed that allows vigorous backwash with minimal loss of media. Engineers usually design the height of a trough to allow for 50% bed expansion without loss of media into the troughs. Troughs may be constructed of concrete or fiberglass and should have smooth, level edges to allow for even carryover. Operators should periodically observe the beginning of the wash cycle for uneven wash-water carryover, which indicates a need for maintenance of the troughs.

Underdrains at the bottom of the filter support and retain filter media, collect filtrate, and evenly distribute wash water. Unfortunately, some old-style underdrains do not always provide a barrier to the finest media in the bed. These small-diameter media can clog the underdrain and lead to uneven filtration or backwashing. These two problems are described here:

If the filter is operated at design loading rates and an area of the underdrain is clogged, unclogged areas must, by definition, be operating at a higher loading rate. This can cause elevated effluent turbidities. For example, if a 10 ft × 10 ft (3 m × 3 m) bed is designed for a maximum loading of 4 gpm/ft^2 (10 m/h), it is accepting 400 gpm (1.8 m^3/min) at design. If 15% of the underdrain system is clogged, only 85 ft^2 (7.7 m^2) is available. The rest of the system is operating at 4.7 gpm/ft^2 (11.8 m/h), or 17.5% over design capacity.

That same condition will cause localized "hot spots" at backwash, as the upward flow of wash water must find its way around the clogged areas. As it does, the velocity increases, which can displace and wash away the media. Also, that portion of the bed above the clogged area may not get washed thoroughly.

For these reasons, some filters are equipped with inspection ports at the underdrain level, and operators should periodically take filters off-line to perform this important task. Operators should also look for deposits of media underneath the underdrain system, which can indicate a compromised underdrain.

GRANULAR BED OPERATION AND OPTIMIZATION

The normal cycle for a rapid sand granular bed filter begins with placing a clean filter into service and follows the steps of filter ripening, effluent production, and subsequent head loss increases.

Eventually, the bed becomes clogged with particulate material, resulting in particle and turbidity breakthrough. At this point, the bed must be cleaned or backwashed. These steps differ in magnitude and frequency from plant to plant, but the process is similar. How these steps are performed can greatly affect finished water quality.

Filter Ripening

When a cleaned filter bed is first put into service, it may not produce effluent of a quality that meets the WTP's goals (Figure 7-8). Particles in the applied influent water need to attach to the grains of sand or anthracite; otherwise, they will pass into the clearwell with the finished water. Properly treated water (see Chapter 5) brings particles that are "sticky," or destabilized, into the filter. These particles attach with a greater efficiency than those that are less sticky. At first, because the bed is so clean, the voids between the grains of sand or anthracite are large, and particles have a greater chance of passing through them. As particles are attracted and attached, the bed usually becomes more efficient at attracting more particles: Each attached particle becomes an attractive site for more particles to come to rest. The voids get smaller, and it becomes easier for the bed to collect more particles. This process is called *ripening*.

This simplified explanation of the ripening process is helpful for operators because it suggests that some steps are under their control. For example, an operator can see that the better the applied water has been conditioned,

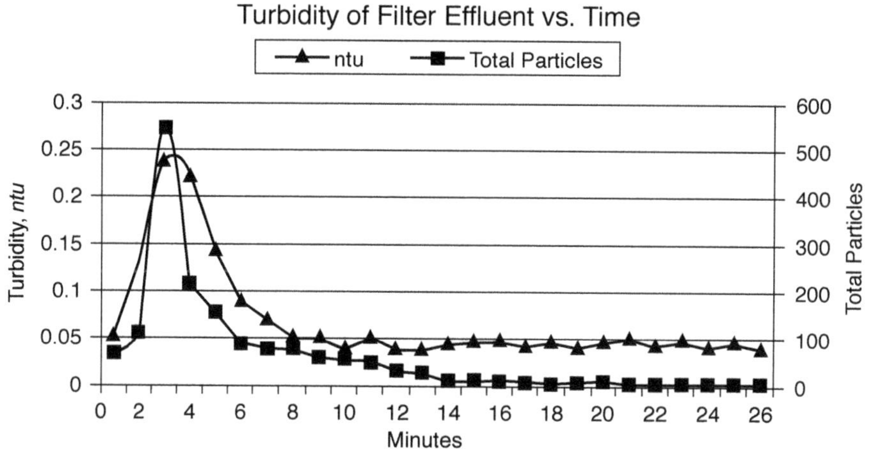

Source: Nick Pizzi

Figure 7-8 Filter ripening graphic plotted with particle count and turbidity data versus time

the better the effluent will be (usually). Also, it is implied that operating filters at slower rates will improve the attachment process because it will reduce the forces caused by drag through the filter. This is called *fluid drag*, and too much of it will shear the attached floc particles. Perhaps the most important lesson is that filters are not just particle-removal devices but are particle-storage devices. Stored particles are in the filter temporarily and must be handled with care.

Head Loss

As the filter ripens and becomes more efficient at producing a clear effluent, head losses increase. Head loss can be defined as a decrease in the available pressure that drives the water through the bed. This loss of pressure is because of frictional pressures brought about by the continual buildup and accumulation (storage) of attached particles. After the filter has been in service for some time, the pressure available from the height of the water in the bed (static head) will equal the frictional pressure, and the rate of filtration will be reduced. At this point, breakthrough may occur faster than particles can attach. Beds should be cleaned before this occurs. Operators usually have a maximum head loss at which they will no longer allow the bed to remain in service.

If filters are allowed to operate at head losses that exceed the static head, a vacuum can result. This is called *negative head* and can cause air binding in the filter—i.e., dissolved gases in the water are released, gas bubbles are trapped, and the problem becomes aggravated. Operators of surface water plants often notice air binding in spring when the air temperature is rising, but the water is still cold. At these times, operators are often tempted to "bump" the filter to release these trapped gases. Bumping is the act of allowing some wash water to travel through the bed in the hope of reducing head loss and increasing the filter run. This practice should be avoided because it will displace particulate contamination into the clearwell. Air-bound filters should be removed from service and backwashed thoroughly, regardless of the length of the run. A better practice is to allow the water level on top of the bed to remain at a height that will minimize head loss increases and therefore minimize air releases.

Rate-of-Flow Control

The ability to control the rate of flow through filters is important (Figure 7-9). Plants with multiple filters may not distribute the total plant flow in a reasonably equal manner because of varying head losses among the filters. Also, without flow control, filters may accept sudden changes (increases) in flow, which can cause poor filtrate quality. In general, flow is controlled in one of two ways: mechanically or nonmechanically (in which hydraulics dictates the flow).

Of course, flow measurement is an integral part of the entire rate-of-flow system. Feedback from the measuring device is important to the filter process. In many cases, it can be used to "pace" chemical feed, such as hypochlorite (OCl^-) or other chemicals.

Source: Lake County, Ohio, Department of Utilities
The signal from this measurement device paces the dosing of hypochlorite to the filter effluent.

Figure 7-9 Rate-of-flow measurement

Rate-of-flow control may be better understood by classifying the filters as either "equal-rate" or "declining-rate" filters. The concept of rate control is complex and will not be discussed in great detail. What follows is a general overview of the subject.

Equal-rate flow control is accomplished by variable-level influent flow splitting, proportional-level influent flow splitting, or proportional-level equal-rate control. None of these conditions can be classified as "constant-rate filtration" because they can occur only when the total plant flow rate and the number of filters in service are held constant. Removing a filter from service usually places a burden on other filters, so constant rate over time is rarely seen.

Variable- and proportional-level flow splitting are attained by placing weirs or orifices on each filter inlet above the filter's maximum water level. The filter effluent discharges to the clearwell at a level above the surface of the filter media. As solids accumulate in the bed, the water level rises in the filter box to provide the head that is needed to drive the flow through the filter. The water level in each box is different and depends on the amount of clogging. When the level reaches a predetermined height, the filter must be backwashed. No instrumentation is used to control flow rate or head loss measurement. Operators can view relative head loss by looking at the level in each box.

Influent flow-splitting weirs are also used for proportional-level operation, which is accomplished mechanically. Each filter requires a level transmitter and controller, with a modulating valve on the effluent. This valve does not control flow—it controls the level on top of the filter. As the water level on the filter changes, the valve attempts to compensate. As head losses accumulate, the valve cannot continue to operate, and so it shuts down, indicating a need for washing.

The most common proportional-level system is the mechanical-control type. This system splits the flow equally among all filters in operation by keeping the influent channel level relatively constant. Effluent valves are modulated to control flow rate, and filters are washed at a predetermined head loss.

Variable- and proportional-level systems are also found in declining-rate filters. Because the influent enters each filter at a level below the normal water level in each filter and discharges to the clearwell at a level above the filter media, the filters operate with the same available head at any instant. The cleanest filters will therefore operate at the highest flows, and the dirty filters will operate at lower flow rates. All filter rates decline in a stepwise fashion after each backwash. When a clean filter is placed into service, it assumes the highest flow. All other filters in service step down from that point.

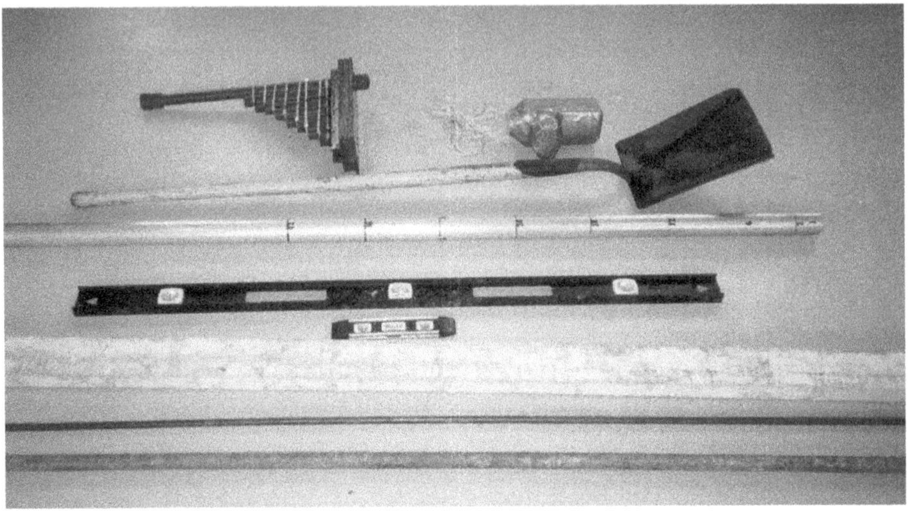

Source: Nick Pizzi

Figure 7-10 Ordinary tools found in most water plants will help operators with basic filter inspections

GRANULAR BED MAINTENANCE

Granular filtration beds tend to be maintenance-intensive. It is common for operators to ignore these important maintenance tasks, which can lead to filter problems that jeopardize filtrate quality. As these problems become more severe, they also become more expensive and difficult to remedy. Most maintenance tasks can be accomplished with the simple tools found in most WTPs.

Bed Depth

Because filter backwashing and other operating practices can result in media loss in the filter, the total media depth should be measured periodically to determine if the existing media still meet original specifications. It is not unusual to lose some media each year, especially if the bed is of dual-media construction composed of sand and anthracite. The backwash velocities required to clean a sand bed may wash away some of the anthracite, which has a density less than that of silica sand. Excessive media loss can, however, indicate poor operational technique or filter bottom problems and will eventually result in filter effluent degradation.

How Deep is Deep?

In part, a filter's ability to trap floc particles in suspension at normal operating rates is a function of bed depth and media size. The minimum standard for proper filter construction is that the ratio of the bed depth, in millimeters, divided by the effective size, in millimeters, is equal to or greater than 1,000, or

$$l/d_e \geq 1{,}000$$

where l = depth of the filter bed, in millimeters, and d_e = effective size of the media.

In practice, it is common for plant operators to maintain a bed that is 10–15% greater than the minimum standard. For example, a 6-ft-deep monomedia bed composed of 1-mm anthracite coal would be more than sufficient because 6 ft is 72 in., and 72 in. × 25.4 mm/in. = about 1,830 mm. Using the formula, 1,830 mm/1 mm = 1,830—a ratio that meets the guidelines for sufficiency.

Beds other than monomedia or dual-media, such as trimedia or coarse monomedia beds, may require a higher ratio.

How Does an Operator Measure Bed Depth?

Bed depth is a measure of the amount of filter media in the cell and does not include the support gravel or underdrains. After checking the original specifications, measure the bed depth by poking a 3/8-in. steel rod into the media until the gravel or support media is reached. Experienced operators develop a "feel" for the difference in resistance that the rod meets at the interface of the media and the gravel. Others can hear the distinctive "crunch" as the rod hits its mark. When the depth has been found, the rod should be pinched at the surface and carefully pulled out. The rod should be taped at that mark and measured in inches. At this point, the rod should be poked into other areas of the filter bed to determine if the filter media is level or the underdrain is disturbed. Within reason, the level should always come close to the tape mark wherever the rod is poked in.

This measurement is very important when discussing bed expansion, and it will quickly indicate when additional filter media need to be installed. Take care when replacing filter media—adding media of a different specification to an existing bed may cause filtering problems over time. Also, be sure to measure the depth of each type of media in dual- and mixed-media beds. It is not uncommon to lose sand in a dual-media bed. Adding anthracite to a bed that has lost sand may worsen the problem.

Bed Expansion Measurement

The cleanliness of a filter bed depends on an operator's ability to achieve and sustain a proper bed expansion during the backwash cycle. This expansion, commonly recorded as percent of bed depth, should remain constant year-round. Water supplies that are subject to wide temperature variations will need to use a different wash-water flow rate to keep this constancy. In general, a higher flow rate is needed during warm water periods than during cold water periods, as warm water has less lifting power than cold water because of the difference in viscosity. Therefore, operators need to make adjustments from one period to the next. Failure to make these adjustments can result in a poorly cleaned bed or a loss of media because of carryover.

Source: Aqua Ohio—Struthers Division

Figure 7-11 Bed expansion device made at Struthers, Ohio, WTP

As an example, consider a filter bed that has a total depth of 30 in. Suppose the operational staff has determined that a 30% bed expansion keeps the bed clean and able to produce the desired effluent turbidity. The calculated bed expansion would then be 30% of 30 in., or 9 in. Any wash-water rate of flow that achieves a 9-in. expansion would be acceptable to the plant staff. The operator simply needs to periodically measure this expansion and adjust the flow rate to maintain the 9-in. expansion.

Operators have devised simple yet effective tools for this purpose, chief of which are the "pan pipe" expansion tool and the circular disc tool. The expansion tool is a series of plastic pipes mounted to a stand at 1-in. intervals. When secured to the top of the bed with cable ties, backwash expansion can be measured by observing the anthracite that has been displaced into the pipes. The highest pipe that has media in it will give an indication of the height of the expansion.

The circular disc tool is simply a white plastic disc fastened to a rod that is marked in inches. The tool is placed so that the disc sits at the top of the bed. A measurement is then taken and recorded. The tool is removed, and the filter is backwashed. At the height of the backwash, the rod is lowered into the washing filter until the disc "disappears" just at the top of the expanded anthracite. A reading in inches is taken at that point. Subtracting the smaller reading from the larger reading gives inches of expansion.

Both methods are usually more reliable than the older method of measuring "rate of rise" by watching the backwash water level travel from one point to another. That method requires the operator to drop the bed level to a predetermined mark and then record the time it takes to reach the next mark. This reading, which is inches of rise, can be converted to gallons per minute per square foot by dividing by 1.6. It does not reveal the extent of bed displacement but rather reveals the backwash water rate.

Filter Coring and Solids Retention

The effectiveness of the backwash scheme and the efficiency of the filtering process can be determined by taking samples, or cores, of the bed at successive depths and measuring solids retention. Filter coring is an inspection technique that examines the amount of floc particles that are attached to the filter media at successive levels for the entire depth of the bed. Solids retention data are expressed as nephelometric turbidity units per 100 g of media. If a solids retention profile is constructed for each filter at a time when the filter is consistently producing acceptable finished water under normal operating conditions, that profile can act as a "baseline," or description of the filter against which other profiles may be compared (see Table 7-2). The table suggests the ranges of floc retention amounts and corresponding actions that might be followed.

Note that the solids retention guidelines are a reference for plants that have not performed this analysis previously. These guidelines may not be appropriate for each filter; therefore, the operator is advised to use these guidelines only as intended. Examples are given in the following paragraphs.

How to Perform a Solids Retention Analysis

The tools for extracting a core sample for solids analysis are easily made at most WTPs. A simple tool made of a 5-ft length of ½-in. electrical conduit is used as the corer. The tool should be marked with an indelible marker at the 2-, 6-, 12-, 18-, 24-, and 30-in. marks, etc. (i.e., for the entire length necessary to reach the bottom of the bed). This configuration of tool material and diameter seems to work best. Other materials (plastic pipe) and diameters do not seem to provide sufficient adhesion for both sand and anthracite, nor do they provide volume sufficient for sample analysis. In addition to the coring tool, sealable plastic bags are needed to hold the media samples.

Table 7-2 Solids retention guidelines for core samples taken after washing

Solids Retained *ntu/100 g media*	Condition of Bed	Action Needed
<30	Too clean	Examine wash rate and length
30–60	Well cleaned and ripened	No action needed
60–120	Slightly dirty	Reschedule retention analysis soon
>120	Dirty	Reevaluate filter wash system and procedures
>300	Mudball problem	Rehabilitate bed

Source: Wolfe and Pizzi 1999

Source: Cleveland, Ohio, Division of Water

NOTE markings on pipe, which correspond to layers in the bed; 1½-in. electrical conduit is used.

Figure 7-12 Operator using coring tool

Clear plastic bags allow for visual inspection of the media, making it possible to observe mudballs and other debris that may be present. A filter core set should be collected from a filter bed after completion of an entire filter run. This set of samples is the "before" set because it is taken before the filter is backwashed. After the filter is backwashed, another set of filter core samples is taken. This is the "after" set because it was taken after completion of a backwash. A before-and-after set of bags should be provided, two each for each stratum (i.e., 0–2 in., 2–6 in., 6–12 in., etc.) that will be sampled. The bags should be coded with the marker to correspond to the levels on the coring tool.

Coring is performed after the bed has been drained. For best results, begin the procedure while the bed is still damp—this will provide for optimal media adhesiveness. Place two small (2 ft × 2 ft) pieces of plywood on the bed so that two operators can stand comfortably on the filter media with minimal bed disruption. One operator inserts the coring tool into the bed at right angles until the 2-in. mark is reached (i.e., to obtain the media sample for the 0–2 in. strata). Carefully extracting the tool so as not to lose the sample, the operator empties the sample from the coring tool (by blowing into the other end of the conduit) into the appropriately marked bag, which is held open by the second operator. This procedure is repeated two more times at two nearby, but separate, locations so that reasonably representative samples of the upper 2 in. of media are obtained.

The next extraction is obtained from the 2- to 6-in. strata, and the operator must carefully insert the coring tool into the same three holes to get these samples. Care must be taken so as not to contaminate the coring tool with material that may have fallen into each sample hole. The tool exterior should be wiped clean between each sampling as an added precaution. This procedure is repeated until samples from all representative strata of the bed are obtained. As each appropriately marked bag is filled, it is sealed and stored for future analysis. For dual- and mixed-media configurations, at least one bag will yield a mixed layer (e.g., sand and anthracite), which should be closely examined for any interesting characteristics, including mudballs, particle size distribution, and, of course, the depth at which it was obtained. The d_{90}/d_{10} ratio of the two media has a direct effect on the length of this stratum.

After all "before" core extractions have been obtained, the bed should be backwashed in the usual fashion. This is an excellent opportunity to observe the backwash process for any anomalies. Additionally, several samples of the backwash water from the point at which the water exits the trough can be obtained for turbidity analysis. A plot of backwash water turbidity versus time can offer additional insights (i.e., at which point during the backwash does the turbidity of the backwash water reach its minimum level?). After backwashing, "after-core" samples of the bed should be obtained by repeating the entire filter-coring procedure.

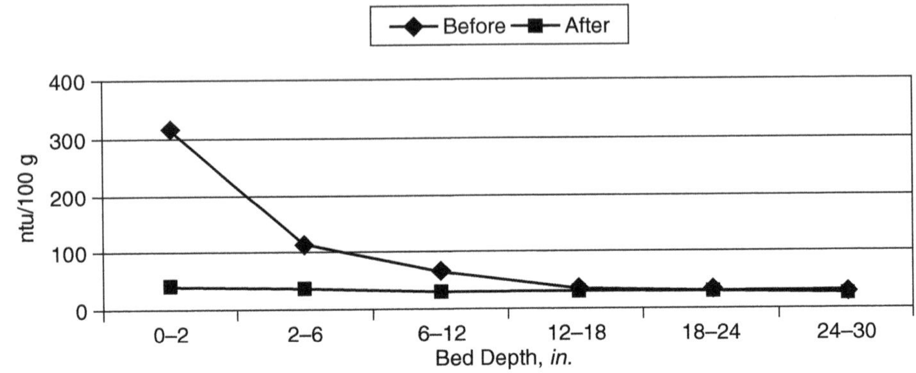

Source: Nick Pizzi

Figure 7-13 Solids retention plotted on a graph

The bags containing the before- and after-core samples are brought to the lab for turbidity, or floc retention, analysis. (This is also a good time to visually check each bag for mudballs.) The goal is to measure floc retention in terms of turbidity, as nephelometric turbidity units per 100 g of media. In the lab, 50-g portions from each bag are weighed and placed into separate 500-mL beakers. Three or more successive 100-mL portions of turbidity-free water are used to wash down and agitate each media sample for 30 s. NOTE: Whatever number of 100-mL portions you start with, you must use the same number of portions for all the samples. After each sample washing, the 100-mL portions are carefully poured into a 1,000-mL beaker, and the turbidity of the resultant suspension is measured. When all measurements are recorded, the results are multiplied by 2 (i.e., to obtain turbidity in terms of nephelometric turbidity units per 100 g of media) and graphed. See Figure 7-13 for an example of the plot.

Graph shows that the filter bed was not overly cleaned. It also shows that solids are not penetrating to the lower levels of the bed, indicating a conservative rate of flow is being used.

As previously mentioned, suggested guidelines for sludge retention profiles for filters are given in Table 7-2. Operators should remember that conditions at their utility (i.e., process type, operating habits, etc.) may produce significant deviation from these published guidelines. Consequently, these guidelines may be inappropriate or misleading at an individual utility and are provided to encourage vigorous examination of local filter operating and backwashing conditions. In general, if filter effluent quality is good, the guidelines in Table 7-2 may be considered irrelevant with respect to local conditions. The operator can simply analyze for solids retention and develop his or her own guidelines as a baseline. If filter effluent quality is poor, these guidelines offer a good starting point and a realistic goal for comparison. Each operator should, in time, be able to develop a "normal" set of guidelines for use based on repeated and regular analysis of the utility's filter conditions.

Filter Backwashing

The most common problems associated with properly designed filters are those induced by the operator. The backwash procedures that operators do or do not follow will determine whether the filter produces quality effluent and can lead to early filter degradation. For this reason, it is important that each plant establish clearly written backwash procedures and that all operators follow them as written. It is not uncommon to see WTP operators in the same plant using different backwash procedures, with all operators convinced that their method is the proper one.

Because there are many design differences among filters and because each WTP is unique, a single backwashing procedure cannot be prescribed. Each plant must develop its own procedures based on design recommendations and experience. The following are some general filter backwash recommendations to be considered at each plant:

> If surface washers are available, use them before the start of the backwash and continue using them as the backwash rate is ramped up to a maximum. Be careful not to allow the surface washers to contribute to media loss.

Surface washing can be a great help in preventing the formation of mudballs. Once mudballs form and continue to grow, they can ruin the ability of a filter to produce low-turbidity water. Mudballs that grow in size also grow in weight and will sink into the filter and go undetected.

It may be advisable to ramp up the wash-water rate in stages; especially if there is more than one type of media in the bed (dual and mixed beds), a ramped wash-water rate that expands each bed properly is advisable. For example, a typical dual-media bed of anthracite and sand requires two rates. The initial rate expands and cleans the anthracite but will not sufficiently clean the sand. However, if the operator ramps quickly to the high rate to clean the sand, the anthracite will be overly expanded and not cleaned properly. Each type of media has a different specific gravity and therefore will require different wash-water rates for expansion. Refer to Table 7-1 for guidelines.

Do not wash the filter too much. Most operators backwash until the turbidity of the wash water coming through the bed is less than 10 ntu or until they can see the surface washers or other piping that is above the bed. Washing the filter too long will remove the necessary bed ripening; the filter will then have to be re-ripened for a longer time before it is useful. This also wastes wash-water pumping. It is normal for operators to want to do a thorough job when they clean the filter, but too much of a good thing can do harm. Operators should periodically collect samples of the backwash water

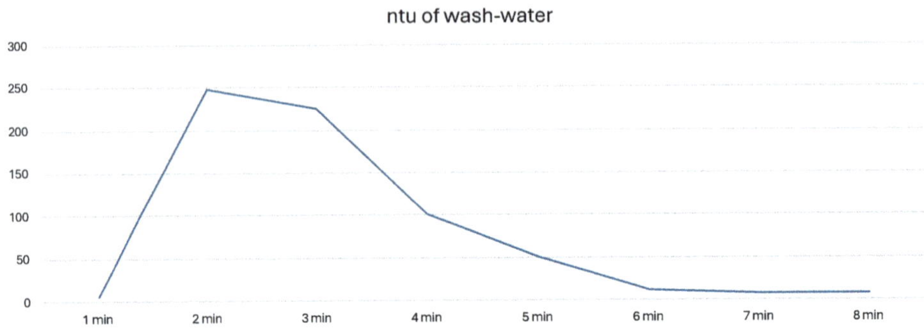

Source: Nick Pizzi

Figure 7-14 Filter wash-water turbidity plotted each minute for about 10 min

at 1-min intervals for the length of the wash and analyze them for turbidity. Figure 7-14 displays filter wash turbidity over time and shows that after 6 min, the backwash no longer produced appreciable solids, yet the operator needlessly continued the wash.

Wash at a rate that will expand the bed properly under all wash-water temperature conditions. Because colder water has more lifting power than warmer water, it is usually necessary to adjust the wash-water rate from summer to winter for most surface water plants in the northern regions (see Figures 7-15 and 7-16 for example wash-water storage tanks). As previously mentioned, the percent bed expansion should not change from season to season, but the wash-water rate needed to achieve the expansion changes with water temperature.

An insufficient wash-water rate will not properly clean a filter no matter how long it is washed. The turbidity of the wash water coming through the filter may lessen, leading the operator to believe that the bed has been cleaned, but this may not be so. This practice usually leads to mudball formation. Washing a filter at 15 gpm/ft^2 for 10 min requires the same amount of water as washing the bed at 7.5 gpm/ft^2 for 20 min, but the results will be different. Filters get cleaned by using the correct amount of wash water at the correct wash-water rate for the correct amount of time. That correctness is best determined at each individual plant using procedures outlined earlier. Table 7-3 shows temperature correction factors for backwash rate values if the rate at 25°C is known.

If air scour is used before (not concurrent with) water backwash, do not assume that less wash-water rate can be used. Air scour (Figure 7-17) will do a good job of knocking the adhered floc particles loose from the filter media, but they still have to be swept up from the bed and into the troughs for removal. This may take as much wash-water rate (gallons per minute per square foot) to accomplish as those filters without air scour; however, it may not take as much wash-water volume.

Figure 7-15 Wash-water elevated storage tank (Lake County, Ohio, Bacon Road WTP)

Source: Aqua Ohio—Struthers Division

Figure 7-16 Wash-water elevated storage tank (Struthers WTP)

Table 7-3 Backwash rate temperature correction factors

Temperature °C	Multiply 25°C Value by
30	1.09
25	1.00
20	0.91
15	0.83
10	0.75
5	0.68

Source: MacPhee and Becker 1999
Example: If 5,000 gpm provides a suitable backwash at 25°C, an operator could expect to achieve the same bed expansion at 10°C by using 3,750 gpm (0.75 × 5,000).

Source: Aqua Ohio—Struthers Division

Figure 7-17 Air control valve for filter air-scour system

A review of operational procedures for backwash at most WTPs that use conventional granular beds reveals that operators usually use 100–150 gal of wash water per square foot of surface media and, on average, use no more than 4% of the plant production to backwash filters.

After backwash, leave the filter off-line or "rest" it if possible. Many operators report that a rested filter ripens more quickly when put on-line than will a filter that is washed and put immediately into service. This practice of resting filters also allows the operational staff to have a fresh filter in reserve that can be put on-line when another filter is removed for washing. This minimizes hydraulic shock in rate-controlled filtration schemes.

An operator cannot perform a proper backwash without visually inspecting the process as it occurs. With newer, sophisticated, and automated backwash systems, it is tempting for the operator to control the sequence from a remote location. However, the operator cannot see and tend to problems that may be occurring at the backwash site. Operators should observe each wash event for uneven surface agitation and expansion, for "hot spots," and for changes in backwash water turbidity. During the wash, the operator should hose down the side walls of the filter box and the associated piping to remove any accumulated scums, which can harbor pathogens.

Think about how the filter wash event will affect all other plant operations such as pumping rates, wash-water return storage availability, maintenance, and any other planned activities. Operators should know in advance when they will be washing a filter and what may or may not happen when they do.

OPERATIONAL CONSIDERATIONS

Through experience gained over time, operators learn techniques that are helpful for operating filters in their plant. What follows are general operational techniques used by most operators.

When filter effluent turbidities begin to increase above what is normal, the problem can usually be traced to one of two areas: the filter(s) or pretreatment. This may seem simple but is often misdiagnosed, and operator actions sometimes worsen the problem. The best-designed and best-operated filters usually will not produce good-quality effluent if pretreatment is poor. It is instinctive for most operators to bring extra filters on-line when the pretreated water has deteriorated, but this may make things worse. Unripened filters, in particular, will not handle poorly treated influent and may actually pass higher turbidities when brought on-line under these conditions. Of course, if the elevated turbidities are due to compromised filters or if the filtration rate is too excessive, it may be wise to use other filters if available.

Operators should not be fooled into thinking that longer filter runs are more efficient. If filter washes are accomplished with 100–200 gal/ft^2/wash (4–8 m^3/m^2/wash) and backwash downtime is about 30 min, no more efficiency will be gained after a UFRV of 5,000 gal/ft^2/run (200 m^3/m^2/run). This is true over a wide range of filtration rates. Longer filter runs will increase the risk that larger amounts of particles will empty into the clearwell if the filter is upset. All other considerations being equal (water quality, filter rate, etc.), the longer the filter run, the larger the amount of particulates being stored in the filter.

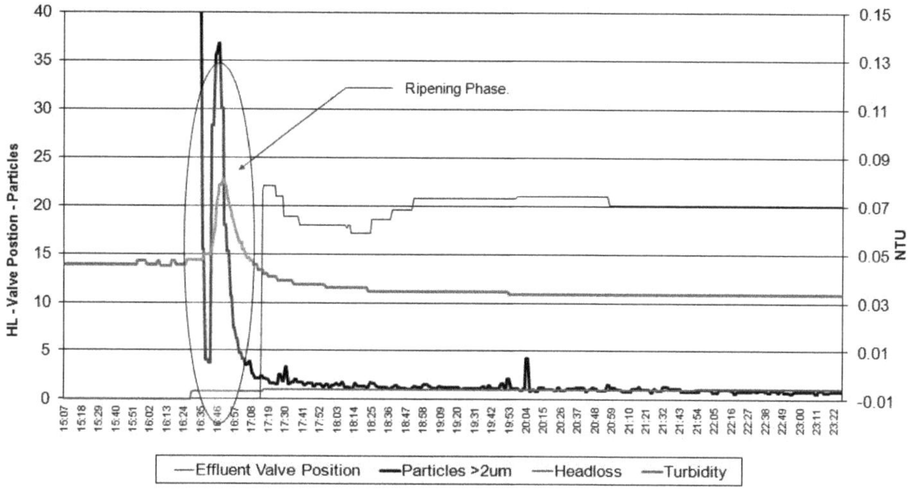

Source: Kinder and Martin 2015

Figure 7-18 Filter run profile highlighting a filter ripening spike

Filter Run Profile

Operators should periodically examine the effluent turbidity or particle count data of each filter and create a graphic profile of this data (Figure 7-18). The profile should show the output of a normal filter run from startup through the end of the cycle. The data will show the ripening characteristics of the filter bed and the effect of hydraulic and treatment changes that occur during the period. Operators should be able to explain any spikes found in the data. These types of profiles are required in the filter exceptions reporting provisions of the Interim Enhanced Surface Water Treatment Rule (IESWTR).

STANDARD OPERATING PROCEDURES

All functions of the filtration process should be documented in standard operating procedures (SOPs) and made available to all operational staff. What follows is an SOP (provided by the Aqua Ohio Struthers WTP) for filter backwash with air scour but no supervisory control and data acquisition (SCADA) capability during an outage.

Backwashing manually during SCADA outage—with and without air scour

SCADA Out of Service (O.O.S).—but manual air scour available

During SCADA failure, operators need to work in tandem to backwash filters as needed. Radios are needed for communications. One operator is stationed at the control in front of the filter as seen on the left. There are four panels located in front of the filters—each panel is set to handle the controls for the two nearest filters. The other operator must station themself by the air-scour control system.

1. Operator #1 switches panel to local (if it is not already).
2. Place air-scour blower in manual.
3. Close influent on the control panel by switching the toggle to close position.
4. Allow filter to drain down through effluent flow to approximately 2–3 in. from bottom of trough. (Water must remain above media—9 in. to anthracite.)
5. Dial effluent to zero (0) using the arrows on red lion pad located on the filter panel.
6. Switch drain to open.
7. Hose down filter walls and troughs.
8. Operator #2 at blower opens master air valve, start blower at 35%.
9. Operator #2 simultaneously closes master air valve as operator #1 opens air-scour valve.
10. Run blower at 35% for 3.5 min.
11. Operator #2 opens master air valve simultaneously as operator #1 closes air-scour valve.
12. Operator #2 stops blower simultaneously as operator #1 opens air relief valve.
13. Operator #1 switches backwash isolation valve to open.
14. Operator #2 switches RTU-B (backwash control panel located near flume) to hand.
15. Operator #2, using the arrows on the red lion control pad, slowly ramps to 1,200 gal/min for 1.5 min (1 min 30 s).
16. Operator #1 observes filter wash.
17. Operator #2, using the arrows on the red lion control pad, slowly ramps to 3,500 gal/min for 4 min in the summer or 3,100 gal/min in the winter.
18. Operator #2, using the arrows on the red lion control pad, slowly ramps down to 1,200 gal/min for 1.5 min (1 min 30 s).
19. Operator #1 switches drain to close on panel (observe filter to normal operating level).
20. Operator #1 closes backwash isolation valve.
21. Operator #2 ramps backwash down to zero (0).

IMPORTANCE OF FILTRATION

With promulgation of the IESWTR, the water quality at a surface water filtration plant is judged by the output of the worst filter on-line. It is no longer determined simply by the quality of the combined filter effluent, which is an average of all filters. This regulatory philosophy returns WTP operations to a time when filtration was the most important unit process available in the quest for public health. Before the advent of chlorination, operators depended on filtration to remove harmful microorganisms from the water supply. Public records of many cities reveal that when filtration was installed at WTPs, a decrease in community disease was noticed. This direct cause-and-effect relationship between unit process and public health was the single most important discovery of the day—a shift in thought that brought science into the craft of water supply and treatment.

Soon after, chlorination was discovered and brought into vogue as a first line of defense against disease-causing organisms. Generations of plant operators lost sight of the importance of filtration and began to think of it as a simple polishing process, one that improved the "clarity" of the water. Improving the clarity, it was said, would discourage consumers from seeking alternative and perhaps less safe water supplies.

Improved detection techniques have shown that protozoa in water supplies can sometimes be quite numerous, both in nature and in the plant process—and in water that is crystal clear. It is known that chlorination cannot inactivate many of these organisms. Other unit processes must be optimized to accomplish that. Optimized filtration, in concert with chlorination, serves as a synergistic and scientific approach to that end and contributes to the overall improvement of community health.

BIBLIOGRAPHY

AWWA. 2011 (6th ed.). *Water Quality and Treatment* (J.K. Edzwald, editor). New York: McGraw-Hill.

AWWA. 2009. M30, *Precoat Filtration*. Denver: AWWA.

Bellamy, W.D., G.P. Silverman, D.W. Hendricks, and G.S. Logsdon. 1985. "Removing *Giardia* Cysts With Slow Sand Filtration." *Journal AWWA*, 77(2):52-60.

Beverly, R.P. 2005. *Filter Troubleshooting and Design*. Denver: AWWA.

Cleasby, J.L., D.J. Hilmoe, C.J. Dimitracopoulos, and L.M. Diaz-Bossio. 1984. "Effective Filtration Methods for Small Water Supplies. Final Report." USEPA Cooperative Agreement CR808837-01-0. NTIS No. PB84-187-905. Cincinnati: USEPA, Municipal Environmental Research Lab.

Kawamura, S. 2000. *Integrated Design of Water Treatment Facilities*. New York: John Wiley & Sons Inc.

Kinder, K. and B. Martin. 2015. *Self-Assessment for Water Treatment Plant Optimization*. Denver: AWWA.

Kirner, J.C., J.D. Littler, and L.A. Angelo. 1978. "A Waterborne Disease Outbreak of Giardiasis in Camus, Wash." *Journal AWWA*, 70(1):41.

MacPhee, M.J. and W. Becker. 1999. "Filter Backwashing: When and How." *Opflow*, 25(3):10-12.

Pizzi, N.G. 2000. "Water Treated Right at Oak Creek Plant." *Opflow*, 26(9):6-7.

Pizzi, N.G. 1996. "Optimizing Your Plant's Filter Performance." *Opflow*, 22(5):1-5.

Pizzi, N.G., R.O. Schwarzwalder, and J.J. Sadzewicz. 1997. "Using Negotiated Guidelines in a Facility Upgrade." *Journal AWWA*, 89(3):89-95.

Tovarnak, J. 2024. Personal communication.

Wolfe, T.A. and N.G. Pizzi. 1999. "Optimizing Filter Performance." *Journal NEWWA*, 113(1):6-21.

Chapter 8

Disinfection

The primary goal of disinfection is to inactivate pathogenic microorganisms in water. Other treatment processes may contribute to pathogen reduction; however, that is not their primary purpose. There are numerous types of disinfection, all of which use chemical or physical agents. The process of disinfection is part of the overall multiple-barrier approach to protecting public health.

WHY DISINFECT?

Many harmful microorganisms, such as bacteria, viruses, and parasites, can be present in water sources. Contaminated water can transmit waterborne diseases, which pose a significant risk to human health. Source water may contain pathogens that are responsible for waterborne diseases, like cholera, typhoid, and dysentery, and must be inactivated. Other processes may be needed in addition to or instead of chemical disinfection because some microorganisms are unaffected by levels that are normally used in the water treatment plant (WTP). Most states, however, require the maintenance of a residual disinfectant such as chlorine (Cl) or chloramine in the distribution system.

CHEMICAL DISINFECTION

There are several methods for chemical disinfection, chlorine being the most widely used. Chemical disinfection by chlorine has existed for more than a century. In early times, chlorine disinfection helped eradicate, or at least greatly reduce, the incidence of classic waterborne diseases such as typhoid and cholera. Disinfection with chlorine is probably one of the most important engineering accomplishments with respect to public health. In recent years, hardier organisms have emerged (e.g., *Cryptosporidium*) that resist traditional disinfection methods, and other technologies to augment or replace chlorine disinfection are being studied. The traditional and the more modern disinfection practices are discussed in this chapter.

BASICS OF CHEMICAL DISINFECTION

There are some fundamental concepts that operators need to know about chemical disinfection. When an oxidant such as chlorine is applied to water, it reacts with many of the impurities in the water to form compounds, some of which do not have the power to disinfect. Because the chlorine is used up in these reactions, it is said that the water has a "demand" for chlorine. Of course, some of the chlorine is used up, destroying the pathogenic organisms, which also contributes to the demand for chlorine. Other impurities that use up chlorine include ammonia, iron (Fe) and manganese, hydrogen sulfide, and organic material, including organic nitrogen. As long as these impurities are found in the source water, some of the disinfectant will be used up in the treatment process. Some reactions, like that of chlorine and ammonia, take place quickly. Others, like the reactions between chlorine and organic material, may take longer.

The oxidation of soluble inorganic contaminants must take place before filtration because the oxidized form of some constituents will stain fixtures if allowed to enter the distribution system. Additionally, all inorganics will create an oxidant demand. Table 8-1 shows the amount of chlorine needed to oxidize some of these inorganic compounds. The process also consumes alkalinity.

Most disinfection strategies require that an excess of disinfectant remain in the water as it travels through the distribution system. This excess, or residual, is desirable because it helps to continue the disinfection process, as well as provides added protection from unforeseen contamination or slime growth in the distribution system. When all of the demand for disinfectant is met, any added amount of chemical will produce this residual.

Effective disinfection takes place when the needed amount of disinfectant comes in contact with the microorganisms in the water and has sufficient time to disrupt the normal life processes of the organism. Pathogens such as viruses and *Giardia lamblia* have minimum requirements for disinfection concentration and contact time, and these requirements are temperature and pH dependent. The concept of disinfection sufficiency is represented by the equation $C \times T$, where C represents a known concentration of disinfectant, in milligrams per liter, that is multiplied by T, the time in minutes. Achieving

Table 8-1 Chlorine requirements and alkalinity consumption with inorganics

Inorganic Reactant	Part of Chlorine Required per Part of Inorganic	Alkalinity Consumed per Part of Chlorine Added
Iron	0.6	0.9
Manganese	1.3	1.5
Nitrite	1.5	1.8
Sulfide to sulfur	2.1	2.6
Sulfur to sulfate	6.2	7.4

Source: Connell 1996

the appropriate $C \times T$ value, in milligram–minutes per Liter (mg-min / L), is a requirement of the Surface Water Treatment Rule (SWTR) and is explained later in this chapter and in Chapter 1.

PRACTICAL ASPECTS OF CHLORINATION

Disinfection using chlorine, or chlorination, is usually accomplished by adding gaseous chlorine, liquid sodium hypochlorite (NaOCl), or solid calcium hypochlorite ($Ca(OCl)_2$). The application of any of these chemicals results in "available chlorine," or hypochlorite (OCl^-), and is represented by the formula OCl^{-1}. Gaseous chlorine from a ton cylinder or a 150-pound cylinder, for example (as shown in Figures 8-1 and 8-2), is considered pure, so 100% of it goes into producing hypochlorite. Liquid sodium hypochlorite is purchased or generated at the site of application. It has a working percentage of 1–16% available hypochlorite, which must be factored by using mathematical formulas such as those found in Appendix B. Granular calcium hypochlorite also is not pure, available chlorine; generally, it is found to be 65–70% calcium hypochlorite. Operators should note the effect that chlorination has on pH. Gaseous chlorine tends to lower pH, while the hypochlorite compounds tend to raise it. The degree to which the pH is affected is dependent on the water's buffering capacity. Low-alkalinity waters can exhibit very noticeable swings in pH when chlorinated.

Source: AWWA 2006

Figure 8-1 Ton cylinder of chlorine gas

Source: Conneaut, Ohio, WTP
Tandem unit allows for dual withdrawal.

Figure 8-2 Installation of 150-lb chlorine cylinders

Properties of Chlorine

Chlorine is a greenish-yellow gas with a penetrating and distinctive odor, and it is very harmful to humans, even at low concentrations in air. The gas is 2.5 times heavier than air. For this reason, it is necessary to move to high ground when a chlorine leak occurs. Chlorine gas has a high coefficient of expansion, meaning that it expands easily and with great volume when heated. This latter property makes it dangerous because it could easily burst its container if heat is applied. Chlorine will not burn, but it will support combustion. Also, water makes chlorine gas corrosive to steel. For these reasons, operators should follow these commonsense rules when using chlorine:

- Always use a self-contained, positive-pressure breathing apparatus when working on chlorine cylinders.
- Never work on cylinders alone.
- Never apply heat to a chlorine cylinder.
- Never apply water to a leaking or burning cylinder because the leak will grow worse.

Chlorine Reactions

When chlorine is added to water, several chemical reactions take place. Some involve the water molecules, and some involve organic and inorganic substances in the water. To study these reactions, it is necessary to become familiar with some terms that are commonly used in the practice of chlorination. Small amounts of chlorine added to water will combine with organic and inorganic materials to form chlorine compounds. As more chlorine is added, a point is reached at which the reaction with these organic and inorganic materials stops. At this point, the *chlorine demand* has been satisfied.

The chemical reactions that took place in satisfying the chlorine demand created compounds, some of which have disinfecting properties and some of which do not. In the same fashion, chlorine will react with water molecules to produce compounds that have disinfecting properties. The total of all the compounds with disinfecting properties, plus any remaining uncombined chlorine, is known as *chlorine residual*. The presence of chlorine residual is measurable and is an indication that chemical reactions have taken place and that there is still some *free* or *combined available residual* to kill the microorganisms that may be introduced to the water.

The amount of chlorine needed should be added together to satisfy the chlorine demand and the amount of chlorine residual needed for disinfection to arrive at the *chlorine dosage*. This is the amount of chlorine that must be added to the water to disinfect it.

Chlorine Dosage Equation

Chlorine Dosage, mg/L = Chlorine Demand, mg/L + Total Chlorine Residual, mg/L

Chlorine Residual Equation

Total Chlorine Residual, mg/L = Combined Chlorine, mg/L + Free Chlorine, mg/L

Reaction with Water

Free chlorine combines with water to form hypochlorous acid (HOCl) and hydrochloric acid (HCl):

$$Cl_2 + H_2O \leftrightarrow HOCl + HCl$$

Depending on the pH of the water, hypochlorous acid may be present in the water as the hydrogen ion and hypochlorite ion dissociate.

Hypochlorous acid dissociation is represented by the following equation:

$$HOCl \leftrightarrow H^+ + OCl^-$$

In dilute solutions of chlorine in water (such as those found in a WTP), the formation of HOCl is complete and leaves little free chlorine (Cl_2). Hypochlorous acid is weak and poorly dissociates at pH levels less than 6. Thus, any free chlorine or hypochlorite (OCl^-) added to the water will immediately form either HOCl or OCl^-; the species formed is thereby controlled by the pH of the water. This is extremely important because HOCl and OCl^- differ in disinfection strength. HOCl has a much greater disinfection potential than OCl^-. Normally, 50% of the chlorine present in water with a pH of 7.3 will be in the form of HOCl, and 50% will be in the form of OCl^-. The higher the pH level, the higher the percentage of OCl^- and therefore the less disinfecting ability.

Chlorination with Hypochlorite

The use of hypochlorite compounds to treat water achieves the same results as chlorine gas in that they each produce HOCl in water as a disinfectant. Hypochlorite may be applied in the form of calcium hypochlorite, $Ca(OCl)_2$ (commonly known as HTH), or sodium hypochlorite, NaOCl (referred to as bleach in liquid form). The chemical reactions of hypochlorite in water are similar to those of chlorine gas.

Calcium Hypochlorite

Calcium Hypochlorite		Water		Hypochlorous Acid		Calcium Hydroxide
$Ca(OCl)_2$	+	$2H_2O$	→	2HOCl	+	$Ca(OH)_2$

Sodium Hypochlorite

Sodium Hypochlorite		Water		Hypochlorous Acid		Sodium Hydroxide
NaOCl	+	H_2O	→	HOCl	+	NaOH

The tank is part of a closed system that pumps hypochlorite solution out to the feeders, which in turn circulates the liquid throughout a network of piping to the feed points. Any excess feed is rerouted back to the day tank (Figure 8-3).

When chlorine gas is used for disinfection, 100% of the gas is available as chlorine. Therefore, if the chlorine demand and the desired residual require 50 lb of chlorine to be fed, the chlorinator would be set at exactly 50 lb/day. However, when hypochlorites are used, the compounds contain only a percentage of the available chlorine that the gas does. HTH is about 65% available chlorine, and sodium hypochlorite is usually found to be 5–15% available chlorine. Because of this, more pounds per day must be fed to obtain the same results that would be expected from chlorine gas. To calculate the requirement (pounds per day) of hypochlorite, first determine

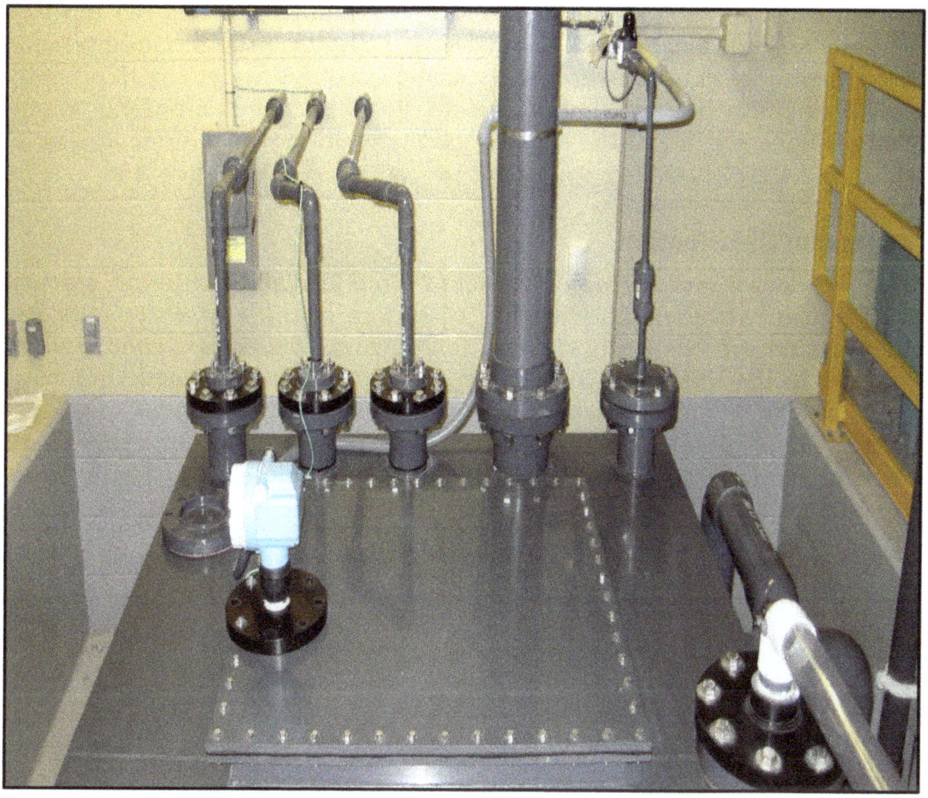

Source: Lake County, Ohio, Department of Utilities

Figure 8-3 Day tank arrangement for sodium hypochlorite feed system

the pounds per day of chlorine gas needed; then, divide that number by the percentage of available chlorine in the hypochlorite. The calculation is:
lb/d Hypochlorite = (Million Gallons per Day)(8.34)(mg/L Gas)/Percent Available Chlorine in Hypochlorite

> *Example 1:* A total chlorine dose of 12 mg/L is needed to treat a source water. If the flow is 1.2 mgd and the hypochlorite has 65% available chlorine, how many pounds per day should be fed?
>
> *Answer:* 12 mg/L × 1.2 mgd × 8.34/0.65 = 185 lb/d HTH
>
> *Example 2:* A flow of 850,000 gpd requires a dose of 25 mg/L chlorine. If sodium hypochlorite is 15% available chlorine, how many pounds per day are needed?
>
> *Answer:* 25 mg/L × 0.85 mgd × 8.34/0.15 = 1,181 lb/d sodium hypochlorite

Normally, these compounds are fed as solutions (Figures 8-3 through 8-5), so the feed (gallons per day) necessary to deliver the required pounds per day must be determined. If the specific gravity is known, multiply it by 8.34 lb/gal and use that number as a divisor of the pounds per day to calculate gallons per day. If the specific gravity is unknown, assume it to be 1, and divide the pounds per day by 8.34 to calculate gallons per day.

Breakpoint Chlorination

Many water systems attempt to produce "free available" chlorine residual in their plant effluent. This process of adding chlorine to water until the chlorine demand has been satisfied is known as ***breakpoint chlorination***. Chlorine in this form has its highest disinfectant ability. Further addition of chlorine will produce a chlorine residual that is directly proportional to the amount of chlorine added beyond the breakpoint. Public water supplies are normally chlorinated past the breakpoint unless the utility practices chloramination.

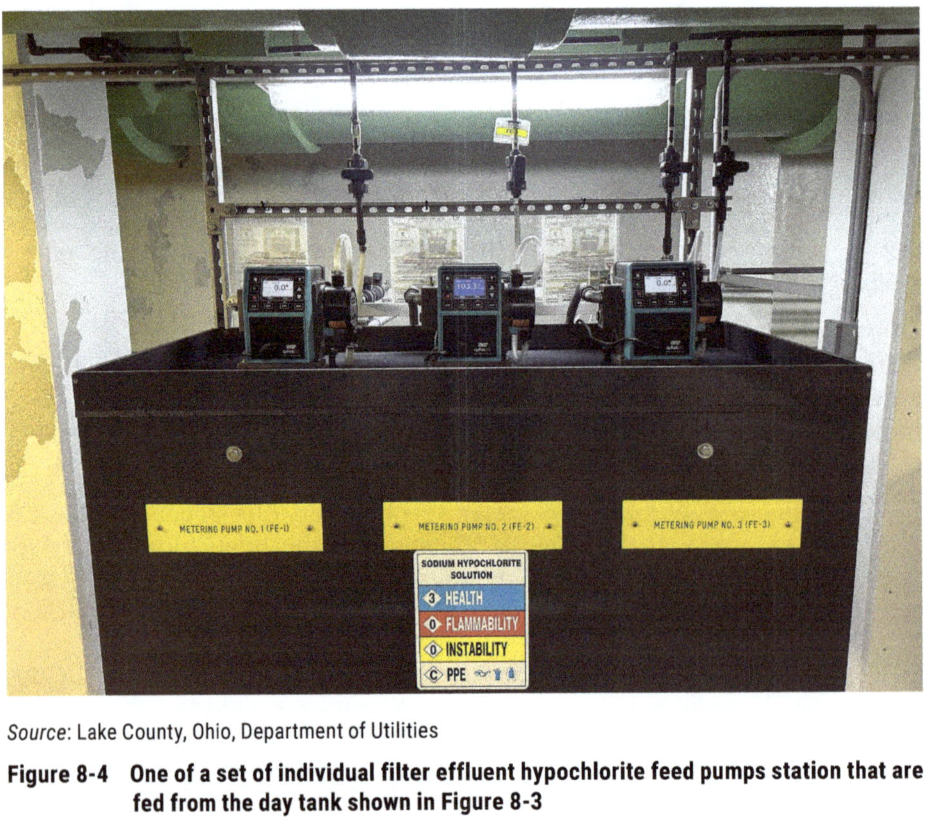

Source: Lake County, Ohio, Department of Utilities

Figure 8-4 One of a set of individual filter effluent hypochlorite feed pumps station that are fed from the day tank shown in Figure 8-3

Source: Aqua Ohio—Struthers Division

Figure 8-5 Day tank at the Struthers, Ohio, WTP

Figure 8-8 is a graph showing the breakpoint chlorination curve. As an example, assume that the water being treated contains some manganese, iron, nitrite, organic matter, and ammonia. When a small amount of chlorine is added, it reacts with (oxidizes) the manganese, iron, and nitrite.

And that is all that happens—no disinfection takes place, and no chlorine residual forms (point 1 to point 2 on the graph). If a little more chlorine is added, enough to react with the organics and the ammonia, chloro-organics and chloramines will form. The chloramines produce combined chlorine residual—a residual that has combined with other components and has lost some of its disinfecting power. These components may cause tastes and odors (from points 1 to 3).

If just a little more chlorine is added, the chloramines are destroyed (points 3–4), and the breakpoint, the point at which a little more chlorine will begin to produce a free available chlorine residual (beyond point 4), has been reached. This chlorine is free in the sense that it has not reacted with anything but is free to do so and will react if anything gets into the water that needs to be disinfected or oxidized further.

Chlorine Feed Equipment

Chlorine can be fed during pretreatment or posttreatment and often is fed during both to accomplish several tasks (Figure 8-6). As discussed in Chapter 5, coagulation is usually enhanced in the presence of a strong oxidant, and chlorine is routinely used for this purpose in the pretreatment scheme. Use of pretreatment chlorine is sometimes minimized if disinfection byproduct (DBP) formation is an issue.

Chlorine gas is fed using a pressure-feeder or vacuum-feeder system. Most often, the vacuum type is used. This equipment operates by sending a vacuum signal to the gas container. A mechanism draws the gas to a stream of water and produces a solution, which is fed at the application point(s). These units are safer than pressure feeders and generally provide more even dosage control.

CHLORINE DIOXIDE

Chlorine dioxide is a greenish-yellow gas at room temperature and is odorous like chlorine. Because it is unstable at high concentrations, it is never shipped in bulk but rather generated on-site. Gaseous chlorine dioxide is generated from solutions of sodium chlorite that have been reacted with chlorine solutions at low pH levels. In solution, the chlorine dioxide remains in its molecular form, ClO_2. The gas formation is shown in the following two reactions:

Using chlorine: $2NaClO_2 + Cl_2(g) \rightarrow 2ClO_2(g) + 2Na^+ + 2Cl^-$

Using hypochlorous acid: $2NaClO_2 + HOCl \rightarrow 2ClO_2(g) + 2Na^+ + Cl^- + OH^-$

Excess chlorine is often used in the generation step to drive the reaction to completion and to avoid the presence of unreacted chlorite. Chlorine dioxide can be dissolved in water and is stable in the absence of light and elevated temperatures. At higher pH and at higher temperatures, chlorine

Source: Aqua Ohio–Struthers Division

Figure 8-6 Set of water softeners used to supply the hypochlorite system at the Struthers, Ohio, WTP

dioxide can revert to chlorite and chlorate, which are undesirable in drinking water. Chlorine dioxide was developed to aid in taste-and-odor control but is also used to destroy iron and manganese. Its use as a primary disinfectant for water supplies has gained popularity because it produces fewer DBPs than chlorine and is effective against *Cryptosporidium*.

Taste-and-odor problems may occur at residuals greater than 0.5 mg/L. Because it introduces chlorite to the water, which can cause certain types of anemia, its use is limited to a US Environmental Protection Agency (USEPA) maximum residual disinfectant level of 0.8 mg/L.

CHLORAMINATION

The practice of chloramination, while not a new technology for disinfection (see Table 8-2), has become popular for water systems that are trying to comply with more stringent DBP regulations. Originally developed to reduce treatment costs and to reduce tastes and odors caused by chlorine, chloramination is now recognized for its ability to prevent DBP formation and to extend chlorine residuals throughout a system. Plants that chloraminate usually add gaseous (or liquid) ammonia at posttreatment (Figure 8-7), after chlorinated water has traveled through a tank or clearwell for primary disinfection ($C \times T$) purposes. This is referred to as *postammoniation*. Many utilities use a chlorine-to-ammonia ratio of 4.5:1 to 5:1 to control nitrification in the distribution system.

The ammonia is supplied either as a liquefied, compressed gas in cylinders, horizontal containers, trucks, and railcars; as liquid aqua ammonia; or as granular ammonium sulfate or ammonium chloride. Ammonia compressed gas is considered 100% strength, while liquid or aqua ammonia is usually 26° Baumé, which is 29.4% ammonia by weight. Ammonium sulfate granules are usually about 21% ammonia-N by weight.

Table 8-2 Utilities with years of experience using chloramination

City	Approximate Start of Chloramination
Denver	1914
Portland, Ore.	1924
St. Louis	1934
Boston	1944
Indianapolis	1954
Minneapolis	1954
Dallas	1959
Kansas City, Mo.	1964
Milwaukee	1964
Jefferson Parish, La.	1964
Philadelphia	1969
Houston	1982
Miami	1982
Orleans Parish, La.	1982

Source: Trussel and Kreft 1984

Source: Aqua Ohio—Struthers Division

Used to create chloramines by reacting with free chlorine residual after disinfection.

Figure 8-7 Ammonia storage and feeders at the Struthers, Ohio, WTP

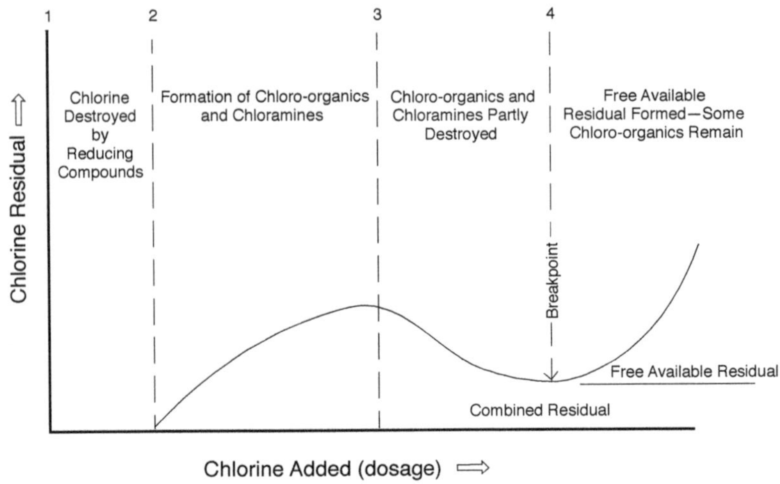

Figure 8-8 Breakpoint chlorination curve

Ammonia cylinders are different than chlorine cylinders in that they tend to be wider and are not equipped with a pressure-relief system. A 1-ton container of ammonia has two fusible plugs and holds 800 lb of material. Because ammonia is fed at about 25–33% of the amount of chlorine used, cylinders rather than larger units (trucks or railcars) are most often found at WTPs.

Ammonia is a dangerous gas, and safety precautions must be taken at all times in areas where it is stored and used. It is readily detectable at concentrations of 20–50 ppm and induces general discomfort at concentrations of 150–200 ppm. It may be fatal at concentrations as low as 2,000 ppm.

Aqua ammonia can be fed from bulk tanks or barrels. Chemical metering pumps are used to deliver the chemical and pace the addition to match the chlorine dosage. The pumps must be equipped with systems to prevent feed malfunctions because of chemical off-gassing. Positioning pumps so that there is a positive suction head usually avoids this problem.

Granular ammonium sulfate can be either fed with dry chemical feeders or mixed with water for solution feed. Depending on the pH of the solution, the solution day tank may need to be sealed to avoid exposure to ammonia gas.

Because the feed of ammonia, as a ratio of the amount of chlorine used, is so important, the ammonia feeder is usually tied directly to the chlorine residual and water flow. Thus, compound-loop control systems are generally used, although some WTPs use manual control of ammonia feed.

The disadvantages of chloramine usage are as follows:

- It is a weak oxidant and disinfectant, especially against certain viruses.
- It may cause taste-and-odor problems.

- Kidney dialysis patients cannot tolerate it and so must be notified of its use.
- It can lead to nitrification problems in the distribution system.

Nitrification is an operational concern, especially in warmer waters. Nitrification is the process by which ammonia present in the water system is oxidized sequentially to form nitrites and then nitrates. If operators do not maintain the proper chlorine-to-ammonia ratio, the excess ammonia can convert to nitrites, and blooms of heterotrophic bacteria can abound. Most studies indicate an optimum chlorine-to-ammonia ratio is between 4.5:1 and 5:1 by weight.

Nitrifying bacteria grow as water temperature increases, and these slimes find refuge in low-flow areas and in tanks with low turnover rates. This causes a loss of residual and creates a higher chlorine demand. Systems then must resort to flushing and tank turnover programs to remediate the problem. Some systems resort to free chlorination once per year to burn out slimes that may accumulate. When this is done, the public must be notified of the change.

The City of St. Louis Water Department uses this strategy to keep the nitrification problem in check. When water temperature increases to greater than 15°C (59°F), the utility will

- maintain a minimum 4:1 chlorine-to-ammonia ratio by weight;
- maintain a finished water chloramine residual concentration equal to or greater than 2.5 mg/L;
- elevate finished water pH to equal to or greater than 9.5 (this is a lime-softening plant);
- increase turnover in the distribution system and reservoir system to shorten residence time; and
- monitor results of heterotrophic plate count bacteria counts using R2A agar.

OZONE

Ozone (O_3) was first used as a disinfectant in the Netherlands in 1893. A few years later, a permanent installation for a continuous ozonator was built in Nice, France. Until the outbreak of *Cryptosporidium* in 1993 in Milwaukee, only a handful of US WTPs used ozone. Since 1993, many utilities have resorted to its use.

Ozone (O_3) must be generated on-site because it is an unstable gas. The formula for the ozone reaction is

$$O_2 + \text{Energy} \rightarrow O + O$$

$$O + O_2 \rightarrow O_3$$

The elemental, or nascent, oxygen (O) produced in the first reaction is provided by an electrical discharge with large amounts of voltage. The air or oxygen that is used must be dry and refrigerated, and it must be particle-free. The yield of ozone can be up to 14% by volume.

Ozone is very unstable in water, and maintaining a residual is difficult. High-alkalinity waters allow for a more stable ozone residual, which is very reactive with natural organic material (NOM) in the source water. When the residual reacts with the organic contaminants, it oxidizes them to a lower-molecular-weight species, including a variety of organic acids. Because these acids may contribute to biofouling of the distribution system, they are usually controlled using biologically active filtration. Ozone may also react with waters containing bromide and thus produce bromoform and other DBPs.

It is difficult to understand or study how ozone attacks and kills pathogens because it is difficult to measure low concentrations of dissolved ozone.

Ozone off-gas collection systems are necessary because ozone is toxic. It must be kept within Occupational Safety and Health Administration limits in the plant and surrounding areas and may be regulated as a discharge gas in some cases. The dissolution of ozone in water produces such a corrosive product that stainless-steel tanks and fluorocarbon gaskets are a must; basins used are made of concrete and caulked with an inert material.

ULTRAVIOLET LIGHT

Unlike ozone, the inactivation of microorganisms by ultraviolet (UV) light is well understood. The reaction is purely physical and therefore not significantly affected by pH. UV light is generated by mercury or antimony vapor lamps and is commonly used to treat wastewater effluent. UV disinfection is often considered for disinfection because of its effectiveness in inactivation of *Cryptosporidium* and because it does not produce DBPs. However, because it does not produce a lasting disinfectant residual, other technologies for residual maintenance (like chlorination or chloramination) are necessary.

UV disinfection is ideally suited to groundwater treatment. The process requires a stable, low-turbidity water. UV does not provide a protective residual, so it is useful in groundwater systems that have a variance that allows them to forego disinfection. UV may then be used as an extra level of protection where residual is not required. Small water systems are also a good fit for UV because most small systems require only a basic level of operator skill to operate.

Process Description

UV light effectively destroys bacteria and viruses. However, how well the UV system works depends on the energy dose that the organism absorbs. An effective dose is measured as a product of the lamp's intensity, including radiation concentration, proper wavelength, exposure time, water quality, flow rate, and the microorganism's type and source, as well as its distance from the light source.

Table 8-3 UV dose (mJ/cm^2) table with inactivation credit

Log Credit	*Cryptosporidium* Dose	*G. lamblia* Dose	Virus Dose
0.5	1.6	1.5	39
1.0	2.5	2.1	58
1.5	3.9	3.0	79
2.0	5.8	5.2	100
2.5	8.5	7.7	121
3.0	12	11	143
3.5	15	15	163
4.0	22	22	186

Source: LT2 provisions of the Surface Water Treatment Rule

Performance

The Long Term 2 Enhanced SWTR lists log credit for UV disinfection of *Cryptosporidium*, *G. lamblia*, and *virus*. These values are listed in Table 8-3. Additionally, UV light disinfection does not form any significant DBPs.

UV effectiveness is relatively insensitive to temperature and pH. UV application does not convert nitrates to nitrites or bromide to bromines or bromates. Studies show that UV-treated drinking water inhibits bacterial growth and replication in the distribution system; however, conditions within distribution systems, such as leaks, still require additional residual disinfection (e.g., free chlorine).

Design and Operation

At a minimum, drinking water systems should install two UV units, both of which should be capable of processing the amount of water the system was designed to handle. Having two units in place ensures continuous disinfection when one unit is being serviced. Two units also can ensure operation during low-flow demand periods. Modular units designed for small drinking water systems are easy to install and operate (two plumbing connections per unit and one electrical hook-up). They should be equipped with automatic cleaners and remote alarm systems. For systems in isolated areas, operators should maintain and store a set of spare parts on-site and consider a telemetry system for monitoring treatment.

$C \times T$ Values

The adequacy of disinfection is determined by the product of the final residual concentration (C) of the disinfectant and the contact time (T) for that disinfectant. The contact time is evaluated as that which is exceeded by 90% of the water—i.e., 10% of the water has less contact time than the T value. These T values are commonly determined by performing tracer studies or by computations based on the ratio of the length and width of the reaction chambers (clearwells or contactors), including any baffling that may be present. Baffling of clearwells, which involves adding separation walls that force snakelike flow patterns, provides greater length-to-width ratios and therefore greater contact times for the disinfectant.

The USEPA has published tables that show the $C \times T$ values needed for basic disinfectants such as chlorine, chloramine, ozone, and chlorine dioxide. $C \times T$ values will vary with temperature, pH, and concentration. Treatment plants for surface water must provide the minimum $C \times T$ value to comply with the SWTR. The $C \times T$ calculation, which is performed every 24 hours, is derived at the highest hourly flow rate and lowest contactor volume for that period. Refer to Appendix B for an example $C \times T$ calculation.

DISINFECTION BYPRODUCTS

There are many DBPs, the chief of which are total trihalomethanes (TTHMs) and haloacetic acids (HAAs). These are formed when chlorine is added to water that contains the necessary types and amounts of NOM.

Chapter 1 discusses the regulatory concerns for DBPs, but operators should understand the kinetics of DBP formation and know which mechanisms are under their control. In general, the following may be considered true for DBP formation:

DBP formation is affected by contact time between the disinfectant and the NOM in water. Therefore, minimizing that contact time, either by adding disinfectant later in the treatment process or by "turning over" the water in distribution storage tanks more frequently, may decrease DBP formation.

The *water age*—the amount of time that water resides in the distribution system or storage tanks before being consumed—is of critical importance. Longer water age usually leads to increased formation of DBPs.

Water age can be estimated by simply performing *tracer studies* of the distribution system. Operators choose representative sample locations in the system and begin sampling them each day for a tracer such as fluoride. The fluoride feed at the plant is turned off (with permission of the regulatory authority), and the time it takes for the fluoride at each sample location to reach background levels will give a good snapshot of relative water age at these locations. Operators can then make improvements by changing valve positions, booster settings, and so forth, which may decrease this age.

Another way to estimate water age is to use a *calibrated* system hydraulic model. Many hydraulic modeling computer programs can generate calculated water age values for locations throughout the distribution system.

Increasing disinfectant dose and residual increases DBP formation. Caution: Operators must avoid the risk of pathogen passage that can accompany disinfectant losses. A balance must be achieved to simultaneously remove microbiological contaminants while controlling DBP formation.

As temperature increases, DBP formation may increase. Warmer water favors the formation of DBPs, as does summer heat, which bakes the system storage tanks.

The water pH can control the type of formation—lower pH favors HAA formation, while higher pH favors TTHM formation.

The nature and concentration of the precursor material will affect the types and amounts of DBP formation. Bromide in the water, or as a contaminant in the chemical feed system, will influence formation of brominated DBP species.

BIBLIOGRAPHY

AWWA. 2011 (6th ed.). *Water Quality and Treatment* (J.K. Edzwald, editor). New York: McGraw-Hill.

AWWA. 2006 (2nd ed.). M20, *Manual of Chlorination and Chloramination Practices and Principles*. Denver: AWWA.

Bolton, J. and C.A. Cotton. 2008. *The Ultraviolet Disinfection Handbook*. Denver: AWWA.

Lauer, W.C., M. Barsotti, and D.K. Hardy. 2009. *Chemical Feed Field Guide for Treatment Plant Operators*. Denver: AWWA.

Trussell, R. and P. Kreft. 1984. Proc. AWWA Seminar on Chloramination for THM Control. Denver: AWWA.

USEPA (US Environmental Protection Agency). 2007. "Simultaneous Compliance Guidance Manual for the Long Term 2 and Stage 2 DBP Rules." EPA 815-R-07-017. Washington, D.C.: USEPA, Office of Water (4601).

USEPA. 2006. "Ultraviolet Disinfection Guidance Manual for the Final Long Term 2 Enhanced Surface Water Treatment Rule." EPA 815-R-06-007. Washington, D.C.: USEPA, Office of Water (4601).

Visintainer, D. 2001. City of Saint Louis Water Commissioner. Internal communication to author. October.

Chapter 9

Softening

The two traditional ways to soften hard water are the lime-soda ash process and the ion exchange process. Membranes, including nanofiltration (NF) and reverse osmosis (RO), can also be used to soften water, and they have gained an increasing use in this field. Membranes are discussed in detail in Chapter 11. Electrodialysis (ED) and ED reversal are other processes that can be used to soften water. They are mostly used for demineralization of high total dissolved solids (TDS) water. A description of the process is included in Chapter 10.

This chapter focuses on the traditional lime-soda process as used at WTPs which treat a wide range of flow rate, such as the 15 mgd Aqua Ohio Massillon WTP in Ohio (Figures 9-1 and 9-2) and the Aqua Ohio—Struthers plant (Figure 9-3).

Figure 9-1 The Aqua, Ohio, Massillon Water Treatment Plant, a 15 mgd groundwater precipitative (lime) softening facility

Source: Aqua, Ohio—Massillon Division

Figure 9-2 Views of the SCUs and solids handling facilities at the Massillon WTP

Source: Aqua Ohio—Struthers Division

Figure 9-3 Soda ash and lime silos, feeder building, wash-water storage tank, and (in the right foreground) three solids-contact units

WATER HARDNESS

As outlined in *Hoover's Water Supply and Treatment* (Pizzi 1995), published by the National Lime Association, the hardness of most water supplies is determined by four compounds that are held in solution. These compounds are

- calcium bicarbonate (solution of limestone, $CaCO_3$, in waters containing carbon dioxide, CO_2);
- magnesium bicarbonate (solution of magnesite, $MgCO_3$, in waters containing carbon dioxide, CO_2);
- calcium sulfate (in the form of $CaSO_4 \cdot 2H_2O$, known as gypsum); and
- magnesium sulfate (in the form of $MgSO_4 \cdot 7H_2O$, known as Epsom salts).

If one or a combination of these compounds is present in a water supply, the water has a degree of hardness. Softening of water requires the reduction or removal of calcium and magnesium from the water—the more that is removed, the softer the water will be.

Calcium and magnesium also form compounds with chlorides and nitrates, which are sometimes present in water supplies and would also contribute to overall hardness. Calcium and magnesium bicarbonates, which were formerly designated as temporary hardness because they could be partially precipitated by boiling the water to drive off carbon dioxide, are now termed *carbonate hardness*. Calcium and magnesium sulfates, chlorides, and nitrates, formerly designated as permanent hardness, are now called *noncarbonate hardness*.

Iron (Fe), manganese, and other divalent cationic ions also cause hardness, but because they are usually present in small quantities, it is customary not to consider them in connection with hardness. However, the presence of iron and manganese is very objectionable (see Chapter 10) because they stain materials with which they come in contact, such as porcelain bathroom fixtures, and are especially objectionable when water is used for laundry purposes. Iron and manganese can be removed separately using oxidation treatment processes, but the softening processes also remove iron and manganese, as well as the other hardness compounds.

TOTAL HARDNESS

The total hardness of water is the sum of the bivalent (two positive charges) metallic cations, primarily calcium and magnesium. Chemical analyses are performed to determine this value. Lab analysts normally determine total hardness and calcium hardness and then subtract the two to get the magnesium hardness. The value is expressed in terms of milligrams per liter as calcium carbonate. For example, if calcium hardness is determined to be 90 mg/L as $CaCO_3$ and the total hardness is 125 mg/L as $CaCO_3$, the magnesium hardness is 35 mg/L magnesium (125–90) as $CaCO_3$. If the values for calcium and magnesium are expressed as the cations, a multiplier can be used to make the conversion so the value can be expressed as $CaCO_3$. These multipliers are derived from the ratio of the atomic weight of $CaCO_3$, which is 100, and the atomic weights of the metals. The atomic weight of calcium is about 40, and magnesium is about 24.3; the multiplier is 2.5 for calcium and 4.1 for magnesium.

Sample Calculation

The lab gives the operator a calcium value of 42 mg/L as calcium and a magnesium value of 12.2 mg/L as magnesium. The operator wishes to express the total hardness as $CaCO_3$.

$$42 \text{ mg/L calcium} \times 2.5 = 105 \text{ mg/L as } CaCO_3$$

$$12.2 \text{ mg/L magnesium} \times 4.1 = 50 \text{ mg/L as } CaCO_3$$

$$\text{Total hardness, as } CaCO_3 = 105 + 50 = 155 \text{ mg/L}$$

Alkalinity

The alkalinity of water is a measure of its carbonate hardness and is determined in the laboratory by titration analysis (see Chapter 12). Carbonate hardness is hardness that is chemically equivalent to the alkalinity, whereas most of the alkalinity in natural waters is caused by carbonate and bicarbonate ions. Noncarbonate hardness is then equal to the hardness minus the alkalinity and represents the calcium and magnesium that are bound up with sulfates, chlorides, and nitrates. In the previous hardness calculation, an alkalinity measurement of 100 mg/L as $CaCO_3$ would yield a noncarbonate hardness of 55 mg/L (155–100).

Sometimes, the alkalinity is greater than the hardness. In those instances, the hardness is equivalent to the carbonate hardness, and the noncarbonate hardness is zero.

The relationship between hardness and alkalinity must be understood to perform the calculation common to the precipitative softening process. For operational purposes, think of alkalinity as existing in one or more of three forms: hydroxide (OH^-), bicarbonate (HCO_3^-), and carbonate (CO_3^{-2}). If any of these three are linked to calcium or magnesium, there is carbonate hardness. The lab test (titration with a standard acid solution to a pH end point of 4.5) that determines total alkalinity (methyl orange test) yields a value, which is labeled as *T*. Another test, the phenolphthalein test (titration to the pH end point of 8.3), yields a value of *P*. The *P* alkalinity is related to the *T* alkalinity in one of five ways, as shown in Table 9-1. The *P* alkalinity can be 0, less than ½ *T*, equal to ½ *T*, greater than ½ *T*, or equal to *T*. The table enables the operator to determine the relationship that the three types of alkalinity have simply by using the *P* and *T* relationship. For example, if the lab shows a *T* alkalinity of 48 mg/L and a *P* alkalinity of 31 mg/L, the relationship of *P* to *T* is that *P* is greater than ½ *T*. As seen in the table, the

Table 9-1 Relationship of *P* and *T* alkalinities

Relationship of *P* Alkalinity to *T* Alkalinity	OH^- Hydroxide alkalinity	CO_3^{-2} Carbonate alkalinity	HCO_3^- Bicarbonatje alkalainity
P = 0	0	0	*T*
P < ½*T*	0	2*P*	*T*–2*P*
P = ½*T*	0	*T*	0
P > ½*T*	2*P*–*T*	2(*T*–*P*)	0
P = *T*	*T*	0	0

NOTE: *P* = phenolphthalein alkalinity and *T* = total or methyl orange alkalinity.
Source: Price 1991.

hydroxides must be 2*P*–*T* or 14; the carbonates must be 2(*T*–*P*) or 34; and the bicarbonate is 0.

Precipitative Softening Process

The precipitative softening process uses chemical reactions that render dissolved calcium and magnesium insoluble. Chemicals such as lime, caustic soda, and soda ash are added to the water in sufficient quantities to create these insoluble compounds. As insoluble compounds, or precipitates, the calcium and magnesium can be settled out and/or filtered, which results in a softer water. The word *precipitate* is used to designate a newly created chemical that has been formed by compounds that were previously dissolved. A precipitate is designated using the symbol ↓. When this symbol follows a chemical compound, it means that the compound is insoluble and will settle out or be filtered under normal operating conditions.

Chemistry of Softening with Lime, Caustic Soda, and Soda Ash

The reactions that take place when lime is added to hard water containing both calcium and magnesium are as follows:

$$Ca(HCO_3)_2 + Ca(OH)_2 \rightarrow 2CaCO_{3\downarrow} + 2H_2O \qquad \text{(Eq 9-1)}$$

$$Mg(HCO_3)_2 + Ca(OH)_2 \rightarrow MgCO_3 + CaCO_{3\downarrow} + 2H_2O \qquad \text{(Eq 9-2)}$$

$$MgCO_3 + Ca(OH)_2 \rightarrow Mg(OH)_{2\downarrow} + CaCO_3 \qquad \text{(Eq 9-3)}$$

$$MgSO_4 + Ca(OH)_2 \rightarrow Mg(OH)_{2\downarrow} + CaSO_4 \qquad \text{(Eq 9-4)}$$

$$MgCl_2 + Ca(OH)_2 \rightarrow Mg(OH)_{2\downarrow} + CaCl_2 \qquad \text{(Eq 9-5)}$$

$$Mg(NO_3)_2 + Ca(OH)_2 \rightarrow Mg(OH)_{2\downarrow} + Ca(NO_3)_2 \qquad \text{(Eq 9-6)}$$

The reactions involving the use of soda ash are as follows:

$$CaSO_4 + Na_2CO_3 \rightarrow CaCO_{3\downarrow} + Na_2SO_4 \qquad \text{(Eq 9-7)}$$

$$CaCl_2 + Na_2CO_3 \rightarrow CaCO_{3\downarrow} + 2NaCl \qquad \text{(Eq 9-8)}$$

$$Ca(NO_3)_2 + Na_2CO_3 \rightarrow CaCO_{3\downarrow} + 2NaNO_3 \qquad \text{(Eq 9-9)}$$

Note that $MgCO_3$ is formed in Eq 9-2, and additional lime is added (Eq 9-3) to convert this $MgCO_3$ to $Mg(OH)_2$. In Eqs 9-4, 9-5, and 9-6, $Mg(OH)_2$ is formed, but the calcium from the lime forms soluble compounds (i.e., $CaSO_4$, $CaCl_2$, and $Ca(NO_3)_2$). Therefore, soda ash must be added (Eqs 9-7, 9-8, and 9-9) to convert these soluble compounds to $CaCO_3$.

The typical water softening plant will analyze the source water for the following four characteristics:

- CO_2 in solution, if present
- alkalinity (total and phenolphthalein)
- total calcium and magnesium (total hardness)
- noncarbonate hardness

From these data, an estimate of the lime and soda ash requirements of the softening process can be produced:

- For aqueous CO_2, lime must be added.
- For calcium carbonate hardness, lime must be added.
- For magnesium carbonate hardness, twice as much lime must be added.
- For calcium noncarbonate hardness, soda ash must be added.
- For magnesium noncarbonate hardness, lime and soda ash must be added.

The balanced chemical reactions can be used to approximate the lime and soda ash required to soften a water completely (i.e., to a minimum achievable total hardness of 40–60 mg/L, as $CaCO_3$). If a higher total hardness is desired in the finished water, less lime or soda ash is added than is required to soften the water completely.

Caustic soda may be used to remove both carbonate and noncarbonate hardness from water. However, when caustic soda is added to water containing soluble calcium and magnesium salts, it precipitates the magnesium as insoluble magnesium hydroxide and absorbs carbonic acid, resulting in the formation of an equivalent amount of sodium carbonate (soda ash). If there is enough noncarbonate hardness in the water being treated to react with the sodium carbonate formed, caustic soda may be used as a softening agent instead of lime and soda ash. As a general rule, caustic soda can replace all the soda ash but only part of the lime. With the higher cost associated with caustic soda, there would be little, if any, savings involved. Laboratory tests under these conditions indicate that the use of caustic soda may leave more calcium carbonate in the settled water than soda ash treatment.

Regardless of whether lime, soda ash, or caustic soda is used to chemically soften hard water, operators should remember that the chemical reactions used to approximate dosages do not account for the purity of the softening chemical being added. The better commercial grades of quicklime are approximately 90–94% CaO, and high-quality calcium hydrated lime is approximately 68–73% CaO. The best grades of soda ash are approximately 99–100% Na_2CO_3, and caustic soda is typically 98–99% NaOH. Therefore, quantities of pure chemicals approximated from calculations using the reactions should be divided by the appropriate percent purity and multiplied by 100 to determine the quantity of commercial-grade chemical that must be added.

Determining Proper Quantity of Lime

If the free CO_2 of the water is known and the magnesium and alkalinity, expressed as $CaCO_3$, are known, the lime requirement can be calculated using the following equation:

$$\left[10.6\left(CO_2\right) + 4.7\left(\text{alkalinity}\right) + 4.7\left(Mg^{+2}\right)\right]/\text{purity of lime, in percent} \qquad \text{(Eq 9-10)}$$

Using Eq 9-10 will give the lime requirement in pounds per million gallons. The constants (10.6 and 4.7) are used to convert milligrams per liter to pounds per million gallons. It is usually necessary to carry an additional or excess amount of lime in the water after treatment, so operators add 10–50 mg/L to the calculated amount. Determining the lime feed requirement in this manner is one of many process control strategies that can be used. For example, some plants attempt to maintain the pH in the softening zone at a specific level. With this approach, the finished water pH varies more than for a plant that controls its processes to maintain a consistent finished water pH by varying the pH in the softening zone.

An example of how one utility establishes its lime dosage is found in the operational techniques at the city of St. Louis, Mo. St. Louis uses a raw river water source, which is highly variable, and consequently bases its lime feed rates on three criteria:

- Maintenance of softening zone pH in the range of 9.8–10.2
- The ratio of phenolphthalein alkalinity (*P*) to total alkalinity (*T*) in the softening zone should not exceed 0.5
- Total river hardness

Generally, to increase the pH 0.3 units as it travels through the treatment process, St. Louis must add an additional 0.3 gpg (5.14 mg/L) of lime. If the *P*/*T* ratio is higher than 0.50, the lime dosage is decreased to lower this ratio. A review of historical data for St. Louis indicates that the optimal *P*/*T* ratio is actually 0.49 in the summer and 0.45 in the winter at one treatment facility.

Changes in total river hardness are also used to anticipate changes in pH and to adjust lime dosages. For every 20-mg/L rise in river hardness, a 0.3-gpg (5.14-mg/L) increase in lime dosage is generally required. It is important to note that such criteria may vary somewhat from plant to plant and between different types of source water.

Recarbonation

After water is softened, it is supersaturated with lime and may need to be recarbonated. Recarbonation is the process of adding CO_2 to the water, usually before filtration, to convert hydroxide and carbonate alkalinity to bicarbonate alkalinity. During the process, the pH will drop to desirable levels. The amount of CO_2 (in pounds per million gallons) necessary for recarbonation can be calculated using this formula:

$$3.67 \left(\text{OH}^- \text{ Change} + \text{CO}_3^{=} \text{ Change}\right) \qquad \text{(Eq 9-11)}$$

where "OH^- change" and "$CO_3^=$ change" are both stated as milligrams per liter. These two components decrease—from before recarbonation to after recarbonation. Table 9-1 can be used to calculate the drop.

For example, if settled water with a *T* alkalinity of 70 mg/L and a *P* alkalinity of 40 mg/L is treated with CO_2 and the operator desires a recarbonated *T* alkalinity of 60 mg/L and a *P* alkalinity of 28 mg/L, how many pounds per million gallons of CO_2 are needed? Using Table 9-1, before recarbonation, $P > ½T$ ($T = 70$ and $P = 40$) and after recarbonation, $P < ½T$ ($T = 60$ and $P = 28$). Therefore, the hydroxides must have been $2P–T$ (80–70), or 10 mg/L before recarbonation. After recarbonation, the hydroxides are zero according to the table. So the total drop in hydroxides is 10–0, or 10 mg/L. The drop in carbonates is also found using the table. Before recarbonation, the carbonates must have been $2(T–P)$, or 2(70–40) = (60 mg/L). After recarbonation, the carbonates drop to $2P$, or 56 mg/L, which is a change of 4 mg/L from before to after.

Using Eq 9-10, the amount of carbon dioxide needed to make this change can be calculated:

$$\text{lb/mil gal CO}_2 = 3.67\ (10 + 4) = 51.38 \text{ lb/mil gal}$$

Amounts of carbon dioxide used will vary from plant to plant and will be affected by water chemistry and the desired finished water pH. In general, low-magnesium waters will not require excess lime for softening and therefore will not require as much recarbonation as high-magnesium waters. Recarbonation can take place in a single stage or in a two-stage process. A single stage is used in low-magnesium waters. Carbon dioxide is fed after sedimentation; operators will normally use enough carbon dioxide to reach a pH of 8.5–9.0 just before filtration. This will dissolve most of the calcium carbonate crystals—the rest are removed through filtration. In the two-stage process (usually used for high-magnesium waters), the source water pH is elevated to 11 or greater with lime. After that stage, carbon dioxide is added to lower the pH to 10.0–10.6, after which the optimal value for calcium carbonate precipitation is reached. After soda ash treatment, more carbon dioxide is added (second stage) to bring the pH to 8.4–8.6. Although operating costs are lower, it is initially expensive because two sets of basins are needed. For this reason, many high-magnesium waters are recarbonated with only one stage.

Carbon dioxide is furnished in pure form as a pressurized liquid by tank trucks, which transfer it to storage tanks (Figure 9-4). In the past, carbon dioxide was generated on-site by burning fuel; this is still done at a few plants. The liquid is vaporized and fed as a gas through a rotameter or sometimes through a chlorinator-type feeder and injected into a flume, which takes the treated water to filtration (Figure 9-5). Operators should use caution in areas where carbon dioxide is stored (Figure 9-6). Although not a poisonous gas, carbon dioxide is colorless and odorless and heavier than air. Without warning, carbon dioxide can displace oxygen (O), and workers can suffocate in minutes.

Source: Aqua Ohio—Struthers Division

Figure 9-4 Carbon dioxide storage tank at Struthers, Ohio, WTP

Source: Aqua Ohio—Massillon Division

Figure 9-5 Recarbonation contact flume at Massillon, Ohio, WTP

Source: Aqua Ohio—Massillon Division

Figure 9-6 Feeder end of carbon dioxide storage tank at Massillon, Ohio, WTP

Plants that soften at relatively low pHs or plants that soften early in the treatment process and then lower the pH to acceptable levels later by adding coagulants, disinfectants, etc., may not need to practice recarbonation. Utilities that do not recarbonate should monitor their filters carefully for carbonate deposition.

Chemical Feed Equipment

In addition to the chemical feeders and pumps that are used in any conventional WTP, a precipitative softening plant needs gravimetric or volumetric feeders of sufficient size to handle the relatively large quantities of chemicals it will use. A glance at Table 9-2 gives the operator a perspective of the amounts of softening chemicals typically used, as compared with other treatment chemicals. This table lists the chemicals used, in pounds per million gallons, in an average day for St. Louis WTPs. Note the difference in quantities. Because more softening chemicals are normally used in a treatment plant than other chemicals, the feed equipment and associated storage facilities are generally quite large and may need more attention and maintenance.

Hardness Tests

The ethylenediaminetetraacetic acid (EDTA) titrimetric test is the test most often used at water treatment laboratories to measure hardness level. It can be performed using standard chemical laboratory methodologies. Alternatively, when ease of use and portability are needed, prepackaged kits can be used. Both methods depend on a titration step. Titration is the process of determining the amount of a substance present in a solution by measuring the amount of a different solution of known strength that must be added to complete a chemical change. In the hardness test, a dye that turns the sample red at pH 10 is added to the water being tested. A solution of EDTA, at known strength, is added to the sample drop-wise until the sample color changes to blue, which is an indication that all calcium and magnesium in the sample has been changed (complexed). The amount of hardness is a multiple of the amount of EDTA used to achieve the change. The operator should refer to *Standard Methods for the Examination of Water and Wastewater* or follow the instructions that come with the kit for proper procedures.

Table 9-2 Relative amounts of chemical used at St. Louis Water Department (compiled from 3rd edition)

Chemical	Pounds used per Million Gallons
Ferric sulfate	200
Chlorine	37
Fluoride	29
Lime	803

Alkalinity Tests

Like the hardness test, the test for alkalinity can be performed with titration methods. Because the total alkalinity test determines the level of hydroxide, carbonate, and bicarbonate activity in the water, it is standard practice to use an acid of known strength as the titrant. A weak solution of sulfuric acid (0.02*N*) is added to the sample that has been previously dosed with a dye. As the color changes, an end point is reached. This occurs at about pH 4.5.

To determine *P* alkalinity, a different color indicator is used. Traditionally, *P* alkalinity is determined by titrating the solution with weak acid until the pH drops to 8.3. The color indication method can be used, or a pH meter can also be used.

Stability Test

Water that is under- or oversaturated with $CaCO_3$ tends to cause corrosion or scaling problems in the distribution system. For this reason, many precipitative softening plants produce treated water with a chemistry that deposits a slight amount of $CaCO_3$ onto the system piping. It is believed to be better to slowly encrust the system rather than corrode the piping with water that is undersaturated.

A stability test can be performed to determine the corrosiveness or scale-forming tendency of water. Two stoppered sample bottles are marked A and B; bottle A is spiked with 0.5 g of $CaCO_3$. Both bottles are then completely filled with water (no air is left in them). The bottles are held at room temperature, and bottle A is occasionally inverted to disperse the $CaCO_3$. Alternatively, a magnetic stirring bar and device are used to keep the $CaCO_3$ in suspension. Over time (1–2 h), the water sample with the $CaCO_3$ either picks up some alkalinity or does not. After the suspension settles, a portion of the settled water from bottle A is filtered and analyzed for alkalinity. A comparison can then be made to the alkalinity of the water in bottle B. If the alkalinity in bottle A is higher than that in bottle B, the water may be corrosive. If the alkalinity in bottle A is lower than that in bottle B, the water may be scale-forming.

Langelier Index

For an indication of a water's corrosivity or scale-forming properties, the saturation index or Langelier index (LSI) should be used.

$$\text{LSI} = \text{pH (measured)} - \text{pHs}$$

For LSI more than 0, water is supersaturated and tends to precipitate (deposit) a scale layer of $CaCO_3$.

For LSI equaling 0, water is saturated (in equilibrium) with $CaCO_3$. A scale layer of $CaCO_3$ is neither precipitated nor dissolved.

For LSI less than 0, water is undersaturated and tends to dissolve (corrosion) solid $CaCO_3$.

In practice, LSI between –0.5 and +0.5 rarely shows corrosion or deposition characteristics.

The pHs (pH at saturation in the previous equation) can be computed using a nomograph and analyses of TDS, pH, calcium, alkalinity, and temperature. This index predicts the final pH that a water should have to be in equilibrium. The reader should refer to *Standard Methods for the Examination of Water and Wastewater* for this test.

Residuals

The residue produced from the precipitative softening process consists mostly of calcium carbonate or a mixture of calcium carbonate and magnesium hydroxide. The sludge solids content is typically near 5%, although a range of 2–30% has been reported. Source waters that require coagulation with aluminum (Al) or iron salts will have residuals that contain those components, as well.

Lime solids can be further thickened and dewatered by facilities such as cake presses (Figure 9-7), centrifuge drying bed, and lime kilns. All are energy intensive, and the operator must weigh the cost of operations against the cost of disposal.

Source: Aqua Ohio–Massillon Division

Residuals are stored and periodically trucked to landfill.

Figure 9-7 Cake press and lime residuals station at Massillon, Ohio, WTP

The amount of dry solids produced during the softening process can be calculated using formulas that account for calcium and magnesium being precipitated and for the amounts of suspended solids removed and chemicals added. The dry weight of sludge solids, ΔS, in kilograms per day, is

$$\Delta S = 86.4Q(2.0\text{Ca} + 2.6\text{ Mg} + 0.44\text{ Al} + 1.9\text{ Fe} + SS + A) \quad \text{(Eq 9-12)}$$

where:

Q = source water flow, m^3/s
Ca = calcium hardness removed, mg/L as $CaCO_3$
Mg = magnesium hardness removed, mg/L as $CaCO_3$
Al = alum dose as 17.1% Al_2O_3
Fe = iron dose as Fe, mg/L
SS = suspended solids concentration in source water, mg/L
A = additional chemicals such as polymer, activated carbon, or clay, mg/L

Ion Exchange Softening Process

Calcium and magnesium can be removed from water by exchanging calcium and magnesium ions for others that do not impart hardness to the water. The ion exchange process (Figure 9-8) involves the use of a resin that has been saturated with nonhardness-producing ions that readily participate in this exchange process. The process used in WTPs is reversible and therefore renewable for many cycles. The method of resin renewal is called *regeneration*. A bed of resin is said to be "exhausted" when it can no longer remove hardness. At that time, regeneration using saltwater (brine) is initiated.

Exchange resins vary in their capacity to soften and be regenerated; it is common to describe them based on the number of bed volumes that can be exchanged before the resin must be regenerated. Also, the amount of calcium and magnesium in the source water dictates the amount of water that can be softened. Harder water exhausts a resin bed sooner than a softer

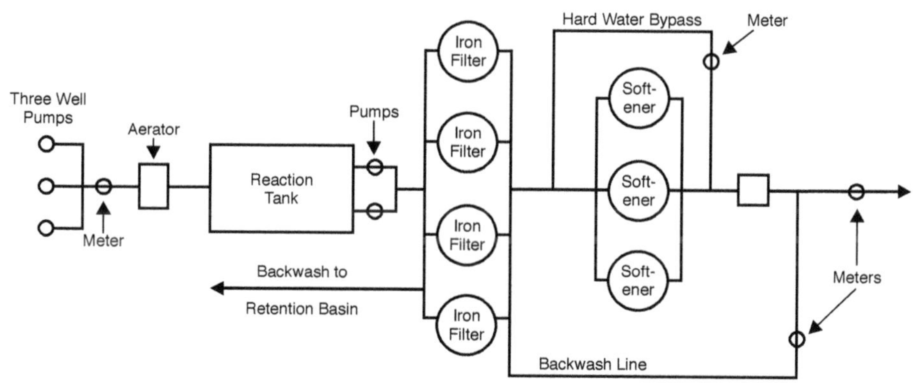

Figure 9-8 Schematic of the ion-exchange process used in the example problems

water. If an operator knows the exchange capacity of the resin bed, the hardness of the water, and the flow rate of water through the bed, he or she can predict the length of a softening cycle. This prediction is an important part of the operational scheme in ion exchange softening plants because it allows operators to schedule time for other tasks and even allows for softening units to operate unattended. More importantly, perhaps, because most ion exchange plants have more than one unit, it allows for staggered operation. In this way, all units are not regenerated at the same time.

Design of Resin Beds and Exchange Units

The hardware involved in the ion exchange process (vessels, piping, valves, etc.) is important; however, it will not be covered in this discussion. Rather, an explanation of the materials used (resin, regenerant) is presented.

Ion Exchange Media

The earliest known materials used to soften water by ion exchange were naturally occurring zeolites found in certain soils. Eventually, synthetic resins were developed that had faster exchange rates and higher capacities to soften. These natural and synthetic materials are rated according to their ability to exchange calcium and magnesium for sodium, which is an ion that does not impart hardness to the water. The rating, or exchange capacity, can be expressed in terms of kilograins (kgr) of hardness per cubic foot of resin. Typically, the modern synthetic resins have exchange capacities of 20–40 kgr/ft^3; natural zeolites were rated at 3–5 kgr/ft^3. Natural zeolites could process about 200,000 bed volumes before replacement was needed. Modern synthetic resins can tolerate about 10 times that number of bed volumes before replacement is indicated.

Multiple terms are used to express hardness, and operators need to know the relationship between these. Hardness expressed as milligrams per liter $CaCO_3$ can be converted to grains per gallon hardness using the following formulas:

$$1 \text{ grain/gal hardness} = 17.1 \text{ mg/L hardness}$$

$$1{,}000 \text{ grains} = 1 \text{ kilograin}$$

An ion exchange unit in a WTP that has a resin rated at 30 kgr/ft^3 can theoretically remove 513,000 mg of hardness (30 × 1,000 × 17.1) for each cubic foot of resin before it is exhausted. In practice, resin beds normally remove a high percentage of the theoretical exchange capacity because of several factors, including flow rates, bed fouling, ionic interference, and efficiency of regenerant distribution. Dead space in the vessel is also a factor because it is nearly impossible to evenly distribute flow through it. However, it is important to realize that until the bed is exhausted, it is considered able to produce water with zero hardness. Beds are usually operated at flow rates of 1–5 gpm/ft^2 (8–40 bed volumes per hour) and will remove 50–75% of the rated hardness-exchange capacity.

Flow Rate Through Bed

Each gallon of water that flows through the bed carries an amount of hardness with it. It is necessary to know the hardness and flow rate values because they will dictate the time of exhaustion. If a bed is loaded at 120 gpm (454 L/min) and the source water has a hardness of 14 gr/gal (240 mg/L), 1,680 gr (120 gpm × 14 gr/gal) of exchange capacity will be used each minute of operation. The operator can see that the bed will exhaust eventually and that the exhaustion time will be reached sooner if the flow rate is increased. Doubling the flow rate will theoretically exhaust the bed in half the time.

As stated, water leaving the ion exchange vessel should have zero hardness. However, because water with zero hardness tends to corrode the system and the customers' plumbing, ion exchange WTPs must produce water that is not of zero hardness. The simplest way to do this is to bypass some of the source water around the softeners and then blend it with the zero-hardness water at a predetermined rate to achieve the desired level of finished water hardness. This predetermined rate is called the *percent bypass* and is calculated by dividing the desired finished water hardness by the source water hardness and converting that value to a percent. This number represents the approximate amount of water that should bypass the softeners. For example, if the source water hardness is 20 gpg (342 mg/L) and the desired finished water hardness is 8 gpg (137 mg/L), the percent bypass is calculated at 8/20, or 40%. If the plant flow is 1,000 gpm (2.2 ft^3/s), the operator knows that 40% of 1,000, or 400 gpm (0.9 ft^3/s), should bypass the softeners to get the desired results.

This bypass water will have the effect of lengthening the life of the softener. More finished water is produced during the cycle because not all of it has to be softened. The greater the desired hardness of finished water, the more efficient the plant becomes with respect to softening expense.

Regeneration

When the bed is exhausted, it must be regenerated. Because the resin has been stripped of sodium ions, these ions must be replaced. This is done by rinsing the bed with a salt solution (brine) at sufficient volume and concentration for a set time. Brine is concentrated sodium chloride (NaCl) solution, which is approximately a 25% concentration at 60°F (15.5°C). The specific gravity of this solution is about 1.192, which yields a working solution of about 2.479 lb of salt per gallon. The brine is usually diluted further before being applied to the softener in the regeneration cycle. As an example, if a softening unit contains 1,000 ft^3 of ion exchange material and it has a hardness-removal capacity of 3,000 gr/ft^3, it will have a softening capacity of 3,000,000 gr. If it is assumed that if 0.5 lb of salt is required per 1,000 gr of hardness removed, 3,000 × 0.5, or 1,500 lb of salt will be required to regenerate the unit. Each gallon of saturated brine contains 2.479 lb of salt; therefore, approximately 605 gal of brine solution will be required to regenerate the unit. To produce a 5% solution (1.03624 specific gravity), the 605 gallons of brine must be diluted to 3,480 gal.

An exhausted resin bed is first backwashed to remove any foreign particles and to restratify the bed. The diluted brine is then introduced through the bed slowly, usually for 20–30 min. The flow rate may be about 1 gpm/ft^2 (0.7 mm/s). It is very important to provide adequate time for the regeneration effect to take place. The bed is then rinsed to remove unused brine before the unit is placed back into service.

Operating an Ion Exchange Plant

Figure 9-8 is a schematic of an ion exchange plant. Note that this plant treats well water that contains iron. Therefore, the plant is equipped with an aerator and iron-removal filters, which is a typical arrangement for ion exchange plants treating water with iron problems (Chapter 10). The iron must be removed before the water is passed through the exchangers; otherwise, the resin will be fouled and rendered useless.

Operators running this type of process must know the design and operating data of the plant. Designers of these plants use a much more complex set of variables, but the following examples represent a basic operational view of ion exchange softening.

There are three ion exchange softeners, and each measures 6 ft (1.8 m) in diameter and has 7 ft (2.1 m) of media depth in them. The exchange capacity of the media is 15,000 gr/ft^3, and it takes 0.3 lb of salt per 1,000 gr of hardness removed to regenerate. The source water hardness is measured at 325 mg/L, and the finished water hardness is 143 mg/L. One 500-gpm (1.1-ft^3/s) high-service pump is operating. With this information, solve the following sample problems.

What is the total hardness-removal capacity of the plant?

The volume of media in the three exchangers is calculated as $\pi R^2 \times$ depth × number of units, or $3.14 \times (3^2)$ ft × 7 ft × 3 units = 593.46 ft^3 of media. At 15,000 gr/ft^3, the hardness-removal capacity would be 593.46 ft^3 × 15,000 = 8,901,900 gr removal capacity. This would also be stated as 8,901.9 kgr removal capacity. There are three units, so each unit has $^1/_3$ of the total capacity, or 2,967,300 gr.

What is the percent bypass?

Percent bypass is calculated by dividing the finished hardness by the source hardness. In this case, it would be 143 mg/L ÷ 325 mg/L, or 0.44. Expressed as percent, this is 44%.

With one high-service pump operating, how many gallons per minute are bypassing the softeners?

One pump operates at 500 gpm; 44% of 500 gpm is 220 gpm. This is the flow rate of the water bypassing the softeners. (This also means that the rest of the water, or 280 gpm, must be going through the softeners.)

How long will one unit operate before it needs to be regenerated?

If 280 gpm is going through the softeners, it can be assumed that the flow through one softener, if balanced, is 280 gpm ÷ 3 = 93.3 gpm. Of course, meters would be installed on the influent to the softeners to help the operator balance the flow, which is desirable. A hardness of 325 mg/L is converted to grains per gallon by dividing it by 17.1. This yields a hardness of 19 gpg. Therefore, a unit is being loaded with source water at a rate of 93.3 gpm, each gallon of which carries with it 19 gr of hardness. After calculating the removal capacity of one unit to be 2,967,300 gr, divide that capacity by 19 gpg to determine that about 156,174 gal of water can be softened before exhaustion. At 93.3 gpm, it will take about 28 hours to exhaust the unit (156,174 gal ÷ 93.3 gpm/60 min/h).

Approximately how much finished water is produced during this time?

This is a tricky but simple question. Regardless of the hardness being removed, if all is going well, the plant is operating at 500 gpm (ignoring backwash, plant use, etc.). Therefore, 500 gpm × 60 min/h × 28 h = 840,000 gal.

Assume the plant is producing 720,000 gal of finished water per day before plant usage. How many pounds of salt will be used in a 30-day month?

Only 56% of that water had to be softened; so in a month, it is calculated (0.56 × 720,000 gal/d × 30 d) that approximately 12.1 mil gal flowed through the exchange units. First calculate the amount of hardness, in grains, that was removed: 12,100,000 gal × 19 gpg = 229,900,000 gr, or 229,900 kgr. Use 0.3 lb of salt per kilograins of hardness removed; so 0.3 lb/kgr × 229,900 kgr = 68,970 lb is needed for the month.

How many gallons of saturated brine is this?

At about 2.5 lb/gal, 27,588 gal (68,970 lb × gal/2.5 lb) would need to be used.

ION EXCHANGE RESIDUALS

Spent brine used in the regeneration process is typically sent to a sanitary sewer. In coastal regions, water utilities may obtain permits to discharge to the ocean. Other methods that have been used to dispose of brine are discharged to evaporation ponds in arid regions or to brine disposal wells if they exist. Disposal of regenerant brine solutions used in cation radionuclide removal must be approved by regulatory agencies.

Other Softening Processes

There are several additional processes that are sometimes used to soften drinking water: ED, membrane softening, and split stream treatment (and blending strategies). ED is discussed in some detail in Chapter 10 because it is used mainly for desalination rather than exclusively for softening. Membrane softening is usually accomplished using specially designed NF membranes (RO membranes also remove hardness). Membrane systems are discussed in detail in Chapter 11. The operation and design of membrane softening systems are analogous to other membrane systems.

Split stream and blending are not separate treatment processes but rather are strategies to adjust water hardness when the opportunity exists. Some treatment processes like RO and ED can remove calcium and magnesium almost completely from the water stream. This may result in a water that is "too soft" from the intended purpose. Some systems blend this demineralized water with unsoftened water to provide a water that meets the hardness goals. This is often termed *split treatment.* Another similar approach is to blend hard water with a softer water source before it is distributed. One example of when this can be accomplished is in a system that has both surface water (soft) and groundwater (hard). The two are blended, and medium hardness water is distributed to customers. When this is possible, this can be a cost-effective approach.

BIBLIOGRAPHY

AWWA. 2011 (6th ed.). *Water Quality and Treatment* (J.K. Edzwald, editor). New York: McGraw-Hill.

AWWA. 2010 (4th ed.). *Principles and Practices of Water Supply Operations—Water Treatment.* Denver: AWWA.

Barnhart, R.K. 1986. *Hammond Barnhart Dictionary of Science.* Union, N.J.: Hammond Inc.

Pizzi, N.G. 1995. *Hoover's Water Supply and Treatment.* National Lime Association. Dubuque, Iowa: Kendall Hunt Publishing.

Price, J.K. 1991. *Applied Math for Water Plant Operators.* Lancaster, PA: Technomic Publishing Inc.

Rodriguez, M.N. 2000. "The Alkalinity Coefficient—A Best Lime Dosage Guide." Unpublished paper contributed by City of St. Louis, Dave Visin tainer memo to author.

Standard Methods for the Examination of Water and Wastewater. 2023 (24th ed.). Washington, D.C.: APHA, AWWA, and WEF.

Chapter 10

Specialized Treatment Processes

This book is purposefully oriented toward conventional water treatment and precipitative softening topics. Most operators are engaged with one or more of those unit processes. However, the author recognizes the importance and growing relevance of specialized processes that help operators meet regulations. Operators are encouraged to seek out the many resources offered by AWWA.

Federal and local regulations and certain site-specific conditions may require the use of specialized technologies for water treatment. The removal of regulated contaminants may require the use of certain targeted treatment processes. Some of these processes may be necessary to produce a product that is aesthetically acceptable to the customer. The descriptions included in this chapter give a brief overview of the most prevalent specialized treatment processes. The references included at the end of the chapter should be consulted for additional information.

AERATION

Aeration, or air-stripping, is used to make water stable and for disinfection, the precipitation of inorganic contaminants, and air-stripping of volatile organic compounds (VOCs) and nuisance-causing dissolved gases. The most used types of aeration are diffused air, surface aerator, spray, and packed-tower systems. These systems remove or minimize the offending contaminants by stripping them out of the water (in the case of gases) or by oxidizing them (e.g., iron [Fe] and manganese [Mn]).

Diffused air or bubble aeration involves contacting gas bubbles with water for the gases to transfer to the water. Operators use this technology to strip VOCs from the water, apply ozone (O_3), oxidize with oxygen, or recarbonate with carbon dioxide (CO_2). The most common arrangement for this process consists of perforated tubes or porous plates situated at the bottom of the reactor tank. This allows for maximum contact time (CT) among the gas bubbles as they rise through the water. The point of application is dependent on the purpose of the process. For example, lime precipitative softening plants that use recarbonation may add CO_2 to the effluent of the clarifiers just before filtration, while operators using ozonation for disinfection may add CO_2 to the source water. The use of ozone technology is covered further in Chapter 8.

Surface aerators are used primarily for oxygen absorption and to strip gases and volatile contaminants.

ADSORPTION (GRANULAR ACTIVATED CARBON)

Adsorption involves the accumulation of a substance at the interface of two phases, such as a solid phase and a liquid phase. An example of this process is the collection of taste-and-odor-causing compounds in water on the surface of activated carbon particles. Ion exchange resins can also be used as interfaces to adsorb organic compounds. The use of ion exchange to remove inorganic compounds is discussed in Chapter 9.

Granular activated carbon (GAC), perhaps the most recognized solid-phase adsorption method used in public water treatment, is a good adsorber because of its surface chemistry and its surface area. GAC particles are honeycombed with cracks and fissures. This property gives the particle a very large surface-area-to-mass ratio. These fissures are capable of capturing (adsorbing) many small organic molecules over extended periods of time. GAC particles, tailored to adsorb different types and sizes of organic molecules, can be specified for different purposes. GAC with small pore openings can be used to trap low molecular weight organics, and the larger pore-sized GAC particles can be used to trap the high molecular weight organics. This same property can, of course, allow the GAC to be fouled or prematurely used up if care is not taken.

GAC is made from raw materials such as lignite, wood, peat, coconut shells, and bituminous coal. The manufacturing process involves heating the material to temperatures near 700°C in the absence of air or to 800–900°C in steam and CO_2. The process and raw materials used determine the characteristics of the final product. GAC is often used as a cap for granular filters and serves the dual purpose of particulate removal and particulate adsorption.

There are several important concepts that operators must know about GAC. Empty bed contact time (EBCT) is a design parameter used to predict performance. It is the detention time of the water that flows through a bed at a given flow rate. The higher the flow rate, the lower the EBCT for a fixed bed depth. For source water of a given quality, there is a minimum depth of GAC that will prevent breakthrough of the adsorbate. All GAC beds remove adsorbates in an area called the *mass transfer zone*. The zone must be deep enough to provide sufficient CT as the water filters down through it. If there is insufficient zone depth or if the design flow rate is exceeded, breakthrough of adsorbate will occur. Therefore, an EBCT that provides adequate opportunity for water/GAC contact is necessary.

Breakthrough will also occur when the bed is exhausted. As the bed removes adsorbate, the available adsorption sites on the GAC become saturated. After a time, the GAC cannot accept any more adsorbate, and it must be replaced or reactivated. For source waters with different kinds of adsorbates, it is common for a GAC bed to become exhausted selectively for each. For example, a GAC bed may begin to pass disinfection byproducts (DBP) long before it passes tastes and odors.

TASTE AND ODOR CONTROL

The control of disagreeable tastes and odors in public water supplies is normally accomplished by oxidation or adsorption processes. In addition, certain odors caused by volatile gases can be stripped from the water by aeration, as seen in Figure 10-1.

There are two strategies that can be used to produce water that is free of tastes and odors. The utility can take measures to prevent or reduce the likelihood that offending compounds will reach the source water, or, once in the water, the utility can take measures to remove them. The choice may depend on economic factors and/or plant capabilities. Most plant operators would agree that prevention is a better practice than removal after the fact, but it is not always practical.

Figure 10-1 Forced draft aerator

The test for taste-and-odor is called the *threshold odor test.* It is an attempt to classify the characteristics, including intensity, of the particular odor that is found in a water supply. The test compares successive dilutions of finished water with odor-free water. The secondary maximum contaminant level (SMCL) for odor is 3 TON (threshold odor number). Operators normally try to keep this value as low as possible because odors tend to erode consumers' confidence in the water supply. When customers lose confidence, they may seek alternative supplies of water that are less reliable or unsafe.

One method for taste-and-odor treatment that is used in conventional treatment plants is to apply a GAC cap to a granular media filter. The detention time is short because the cap is usually only 6–24 in. deep. In a dual-media or multimedia filter, GAC can be added to the top of the filter, or it can replace the anthracite layer. Filter flow rates and head loss characteristics must be considered. Several treatment plants have reported a successful implementation of this approach. Some have used this process to remove DBPs or DBP precursors to acceptable levels.

IRON AND MANGANESE REMOVAL

Iron and manganese are naturally occurring minerals found in rocks and soil that are usually in an insoluble form. However, as rainwater percolates into soils, dissolved oxygen (DO) is removed by the decomposition of organic materials. As oxygen levels decrease, the water becomes capable of dissolving iron and manganese. This is why many groundwaters have iron. Manganese is found more rarely, but when it is present, it is usually with iron.

Iron and manganese do not present public health concerns. They do, however, create considerable aesthetic and operational problems when present in water supplies. As soluble iron and manganese become oxidized, they transform from colorless compounds to turbid yellow and black suspended solids that can stain sinks and fixtures and clog water mains. In some cases, iron or manganese will cause odor problems when they react with other chemicals. Water treatment plants, the sources of which contain these minerals, are usually equipped with treatment devices that remove the minerals before water is distributed to the public.

Because iron and manganese can cause problems, the US Environmental Protection Agency (USEPA) has set the SMCL for iron at 0.3 mg/L. Customers seem to notice a taste from iron if the level reaches 0.5 mg/L. An SMCL has also been set for manganese at 0.05 mg/L. Manganese stains on laundry are very stubborn, and most operators strive to eliminate manganese altogether.

Iron and manganese in water are controlled at the treatment plant using the following methods.

PRECIPITATION AND FILTRATION, ION EXCHANGE, AND SEQUESTRATION

As mentioned in Chapter 9, the lime-softening process also removes iron and manganese quite effectively. However, this process is not solely used for iron and manganese treatment but in combination with softening.

Precipitation Processes for Iron and Manganese Removal

In precipitation, soluble forms of iron and manganese are oxidized to convert them to insoluble forms. When completely converted to an insoluble form, they can be effectively settled and filtered. To be oxidized, the iron must come into contact with air or oxygen, or the iron and manganese must be chemically oxidized using chlorine (Cl), ozone, potassium permanganate ($KMnO_4$), or chlorine dioxide.

Aeration oxidizes iron in water that has a pH greater than 6.5, and aeration is also useful for manganese oxidation when the pH is less than 9.5. Long CTs may be necessary to complete the process. Water can be aerated by forcing compressed air through it or by allowing water to cascade down over trays, effectively allowing for thin sheets of water to be exposed to the air. The trays become coated with iron and manganese hydroxide over time, which promotes further removal of the metals. A reaction tank or vessel is usually provided downstream of the aeration equipment. This tank is designed for a suitable detention period so that complete oxidation of the metals can take place. Detention times of 20 min to 1 hour are common.

Chemicals are commonly used for oxidation and can eliminate the need for double pumping and for the reaction vessel. Chlorine, chlorine dioxide, and potassium permanganate are the oxidants of choice. These powerful chemicals can be used to produce insoluble forms of iron and manganese, and, in the case of potassium permanganate, can reduce the disinfectant/DBP levels because no chlorine is added at the front of the plant.

All these processes require the removal of the iron and manganese floc that is formed. At higher concentrations of the metals, sedimentation is specified, followed by filtration. If pH in the sedimentation basin is low, lime may be used to raise the pH to levels that are more favorable for precipitation. Filtration alone can be used after oxidation if the concentration of the metals is less than about 5 mg/L.

Ion Exchange

Ion exchangers can remove iron and manganese, along with other metals, if the proper exchange resin is used. Where hardness is low and there are no precipitates to foul the bed, this process is specified for waters with low iron content. Otherwise, the process is usually not economical.

Manganese Greensand Filtration

This process is commonly used for the removal of iron, manganese, and hydrogen sulfide from groundwater. The manganese greensand (glauconite) is a naturally occurring material that can be classified to produce an excellent filtration media having a sieve analysis of 18 × 60 mesh with a resulting effective size of 0.3–0.35 mm and a uniformity coefficient of 1.60 or less.

To optimize the process, the material is first coated with manganese oxide. This stabilizes the material and imparts a special oxidation–reduction property that is effective for removing iron, manganese, and small amounts of hydrogen sulfide. The filter media must be regenerated when their capacity to remove iron and manganese is reached. Regeneration is accomplished as either continuous regeneration (CR) or intermittent regeneration (IR).

CR is usually used when iron is mostly being removed. An oxidant, either chlorine or potassium permanganate ($KMnO_4$), is fed continuously ahead of the greensand filters. The oxidized iron is captured on the filter sand. IR is mainly used for manganese removal. The greensand is treated with oxidant (usually permanganate) when the filter capacity is reached. This intermittent treatment occurs while the filter is off-line.

Operating conditions may vary, and optimization is site specific. However, as a starting point, the pH is usually adjusted to be in the 6.2–6.8 range. Alkalinity is ideally greater than 120 mg/L as calcium carbonate ($CaCO_3$). When iron alone is treated, chlorine is fed at about 1.0–1.2 times the iron concentration. For water containing both manganese and iron, most plants feed chlorine followed by permanganate. Chlorine is fed, in this case, at about 1.0 times the iron concentration and 2.0 times the manganese concentration. This is followed by a small permanganate feed that is controlled by the faint pink color in the water just before the greensand filter.

Backwashing the greensand filters must be performed periodically to remove captured iron and manganese particles and to restore the filter rate capacity. To enhance filter performance and improve backwash efficiency, the greensand filters are often capped with anthracite filter media. It is common for the filters to have an 18-in. anthracite cap over 18–30 in. of greensand (depending on the regeneration method and the amount of iron and manganese being removed).

Operating a greensand process can be tricky, and often, careful monitoring of pH, chlorine, permanganate, iron, manganese, and hydrogen sulfide levels is necessary. Sufficient training and instruction regarding proper operation of the chemical feed systems and filter operation are necessary to achieve optimal results.

Sequestration

If the source water is low in iron and manganese (less than 1.0 mg/L combined) and contains no DO, sequestration may be used to treat for iron and manganese. Sequestration is the process by which iron and manganese are kept in solution by adding chemicals, such as polyphosphates or silicates.

In this way, the iron and manganese are not removed, but rather, prevented from being oxidized and made insoluble. Bacterial slimes sometimes occur in systems that sequester because the iron is still available. Care must be taken to maintain adequate chlorine residual to minimize this problem.

FLUORIDATION OF WATER

Fluoride is a mineral that occurs naturally in all water sources, even in the oceans. It is never encountered in the free state but rather in combined form.

When fluoride levels in water are held constant at about 1.0 mg/L, the incidence of dental cavities is reduced. Because of this fact, public water supplies are usually fluoridated to optimal levels. In 2006, the Centers for Disease Control and Prevention (CDC) in Atlanta, Ga., stated that 184 million Americans, or 69% of those served by public water supplies, consume fluoridated water.

Dental caries, the most prevalent chronic disease in human beings, is most widespread in children. Tooth decay reaches a peak in adolescence and diminishes in adulthood. It is for this reason that fluoridated water has the greatest effect on the teeth during the formative years. Fluoride is absorbed into the teeth and bones and is incorporated into the crystal of the tooth enamel. When teeth are forming, this helps to strengthen the enamel by forming a layer that is more impervious to the bacteria that cause decay.

Fluoride levels greater than the optimum can cause mottling of teeth from dental fluorosis. This condition, which creates brittleness in the teeth and moderate to severe discoloration, was first noticed in 1901 in Naples, Italy, as reported by the US Public Health Service. Water purveyors must control the levels of fluoride ion concentration in their supply. The USEPA has set a maximum contaminant level for fluoride at 4 mg/L to prevent crippling skeletal fluorosis. The Department of Health and Human Services and USEPA have recommended that the optimal level of fluoride in drinking water is 0.7 mg/L.

In some communities, there may be controversy surrounding the practice of fluoridation. Most water plant operators are in compliance as directed by local health officials. Assistance with responses to community concerns can be obtained from the CDC.

Chemicals Used in Fluoridation

Fluoride is the negative ionic form of fluorine, which is a gaseous element. Fluoride compounds can be found in all soils, water supplies, plants, and animals in varying amounts. Fluorine is a reactive element and therefore is always found in combination form and never in the free state. Fluoride-containing minerals are used by industry to produce fluoride compounds, the most prevalent being fluorspar, cryolite, and apatite. From these minerals, the three most common fluoride-adjustment compounds are generated: sodium fluoride (NaF), hydrofluosilicic acid (H_2SiF_6), and sodium silicofluoride (Na_2SiF_6). These three compounds are byproducts of the phosphoric acid fertilizer industry, and their availability is dictated by agricultural production. Sales of fertilizer have a direct effect on the manufacture and availability of fluorides for water treatment.

Sodium fluoride

Sodium fluoride (NaF) is a white odorless powder or crystal. It has a specific gravity of 2.79, a molecular weight of 42 (sodium = 23, fluorine = 19), and a solubility of 4 g/100 mL or 4% in water. When added to water, NaF dissociates into sodium and the needed fluoride ions. NaF can be purchased in quantities that are very pure, usually near 97–98%.

A dosage of 19 lb NaF into 1 mil gal of fluoride-free water will produce a 1-mg/L F^- solution. The percentage of available fluoride in NaF is 45% (19 ÷ 42).

The calculation for estimating the dosage of 1 mg/L F^- when using NaF (97% purity) is as follows:

$$19 \text{ lb/mil gal NaF} \times ([1 \text{ mg/L}]/8.34 \text{ lb/mil gal}) \times 0.97 \text{ purity} = 1.0 \text{ mg/L } F^-$$

Sodium silicofluoride

Sodium silicofluoride (Na_2SiF_6) is a white odorless crystalline material. It has a specific gravity of 2.679, a molecular weight of 188 (2 sodiums = 46, 1 silicon = 28, 6 fluorines = 114), and a solubility ranging from 0.44 to about 2.5%. Na_2SiF_6 can be purchased in purities of 98% or higher.

A dosage of 14 lb Na_2SiF_6 into 1 mil gal of fluoride-free water will produce a 1 mg/L F^- solution. The percentage of available fluoride in Na_2SiF_6 is about 61% (114 ÷ 188).

The calculation for estimating the dosage of 1 mg/L F^- when using Na_2SiF_6 (98% purity) is as follows:

$$14 \text{ lb/mil gal } Na_2SiF_6 \times (1 \text{ mg/L} \div 8.34 \text{ lb/mil gal}) \times 0.61 \text{ available } F^- \times 0.98 \text{ purity} = 1.0 \text{ mg/L } F^-$$

Hydrofluosilicic acid

Hydrofluosilicic acid (H_2SiF_6) is a straw- or amber-colored, transparent, fuming, corrosive liquid. It has a molecular weight of about 144 (2 hydrogens = 2, 1 silicon = 28, 6 fluorines = 114) and can be purchased as a liquid in drums or bulk at 23–35%.

A dosage of 46 lb of 23% H_2SiF_6 into 1 mil gal of fluoride-free water will produce a 1-mg/L F^- solution. The percentage of available fluoride in H_2SiF_6 is about 79% (114 ÷ 144).

The calculation for estimating the dosage of 1 mg/L F^- when using H_2SiF_6 (23% purity) is as follows:

$$46 \text{ lb } (H_2SiF_6) \times 1 \text{ mg/L} \times 0.79 \text{ available } F^- \times .23 \text{ purity}/8.34 \text{ lb/mil gal} = 1.0 \text{ mg/L } F^-$$

Note that water has no "fluoride demand," so a contact period, such as for chlorine, is not needed. Some fluoride ions will, however, be lost in sedimentation and filtration (usually 10–30%), so the optimal feed point for fluoride is at postfiltration. The fluoride injection point should be as far away as possible from any chemical feeds that contain calcium, such as lime. It is prudent to check the fluoride concentration on plant tap water daily and to check the feed system hourly while in operation.

Any overfeed of fluoride is a public health concern. All medical and emergency agencies should be notified, especially kidney dialysis centers, if overfluoridation occurs. In addition, the CDC should be notified.

FEED EQUIPMENT

There are three accepted methods for feeding supplemental fluoride to a water supply:

- Fluoride saturator
- Dry chemical feeder
- Chemical solution feeder using either H_2SiF_6 or a solution made from the dry chemicals

Feed equipment is selected based on many factors, including plant size, availability and type of fluoride source, cost, and technical expertise of personnel. Larger facilities seem to prefer liquid storage and feed systems. The operator is directed to AWWA Manual M4, *Water Fluoridation Principles and Practices*, for more information, from which Figures 10-2, 10-3, and 10-4 are taken.

Fluoride chemicals are added to water as liquids but may be measured as either liquid or solid. When solid chemicals are dissolved into solution

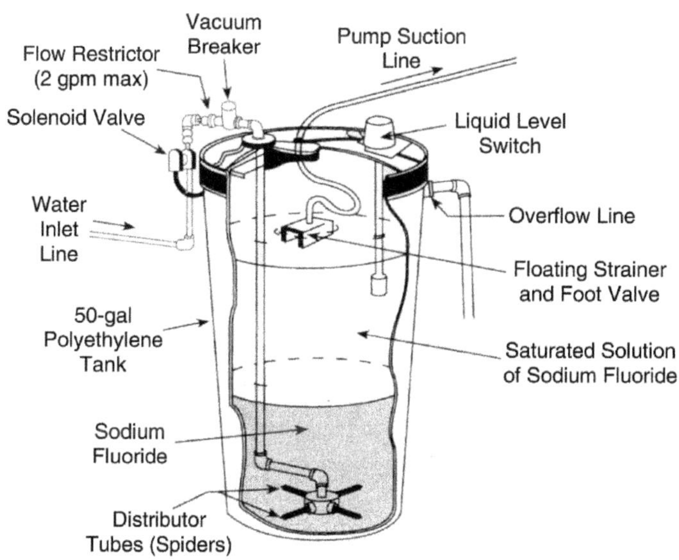

Figure 10-2 Solution fluoride saturator

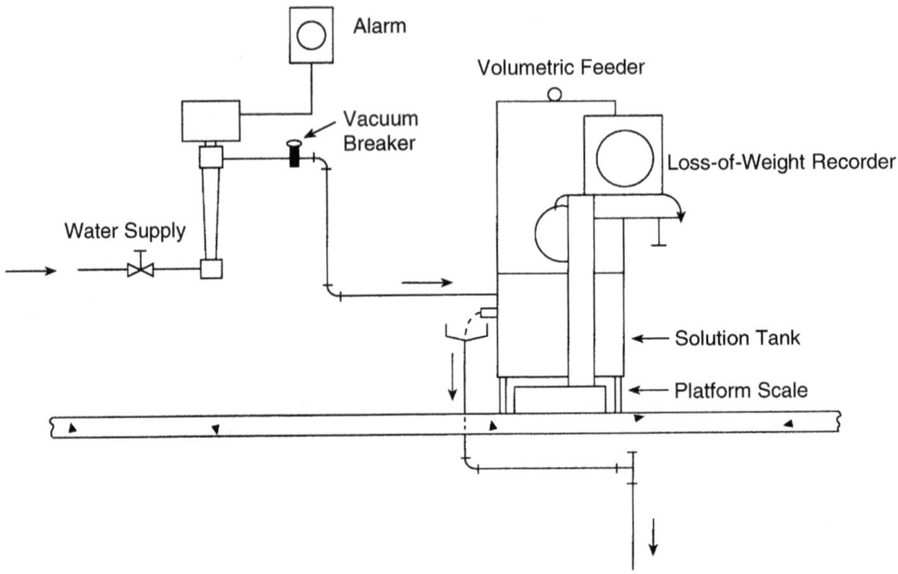

Figure 10-3 Dry feeder installation with volumetric feeder for fluoride

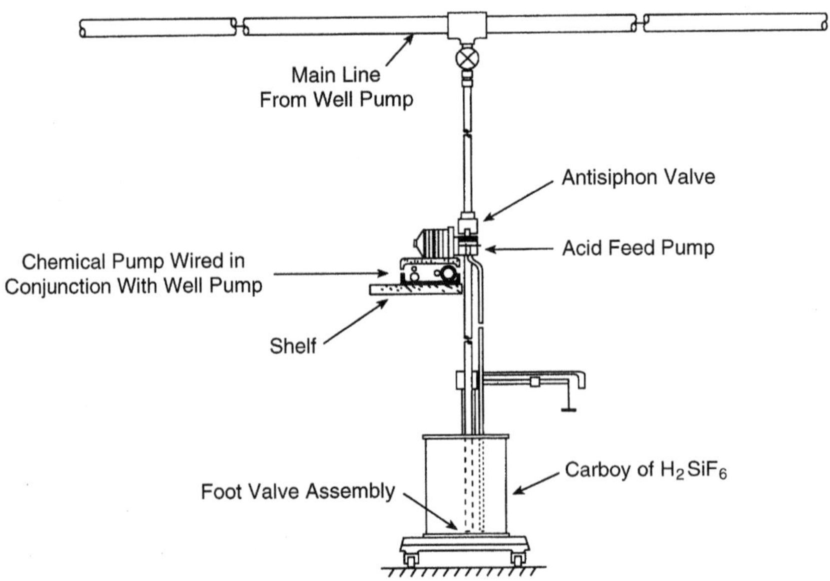

Figure 10-4 Acid feed installation for fluoride

before feeding, the strength of the solution must be controlled, usually by feeding a predetermined amount of chemical for a specific duration. Two types of dry-feed machines exist: volumetric and gravimetric. The volumetric feeder delivers a predetermined volume of dry chemical within a given time; the gravimetric feeder delivers a predetermined weight of chemical within a given time and is more accurate.

SAFETY PRECAUTIONS

Fluoride dust from NaF and Na_2SiF_6 can expose operators to higher-than-optimal levels of fluoride and can irritate the nasal and mucous membranes. Fluoride acids such as H_2SiF_6 can burn the skin. The following precautions are advisable when handling these chemicals:

- Avoid breathing fluoride dust. Dust masks should be issued to all who work around fluoride dust.
- Wear rubber gloves and aprons when handling dust.
- Wash away any dust from the skin immediately.
- When handling hydrofluosilicic acid, wear acid-proof gloves, an apron, and a face shield. Wash these items after use.
- Do not allow people with open cuts or sores to handle any fluoride acids or dusts.
- Post safety rules in the vicinity of the chemical storage and handling area.
- Label fluoride containers as poisonous, and never reuse the containers for other purposes.
- If accidentally ingested, give the victim a 1% solution of calcium chloride and a glass of saturated lime ($Ca(OH)_2$) water or milk to drink. Flush the chemical out of the eyes with warm water. Get medical attention immediately.

EXAMPLE DOSAGE CALCULATION

Problem: A water plant treats 7.21 mgd and wishes to maintain a fluoride ion level of 1.1 mg/L in the plant effluent. If the source water contains 0.36 mg/L natural fluoride ion, how many gallons of 24.3% H_2SiF_6 will be needed in the month of September? The acid used is 10.5 lb/gal.

Answer: H_2SiF_6 has 79% available fluoride ion.

$$1.1 \text{ mg/L desired} - 0.36 \text{ mg/L available} = 0.74 \text{ mg/L } F^- \text{ needed dosage}$$

$$0.74 \text{ mg/L } F^- \times (8.34 \text{ lb/mil gal } F^-/1 \text{ mg/L}) \times 7.21 \text{ mgd} \times (30 \text{ days/month}) = 1{,}334.92 \text{ lb } F^-/\text{month}$$

$$1{,}334.92 \text{ lb } F^-/\text{month} \div (.243 \times .79) = 6{,}953.8 \text{ lb } H_2SiF_6/\text{month}$$

$$(6{,}953.8 \text{ lb } H_2SiF_6/\text{month}) \times (1 \text{ gal}/10.5 \text{ lb}) = 662.26 \text{ gal/month}$$

ACTIVATED ALUMINA FLUORIDE REMOVAL PROCESS

Some source waters contain fluoride in excess of the regulatory limits and thus must be reduced. Both surface and groundwaters may be subject to fluoride removal necessity. Fluoride can be removed by several processes, including reverse osmosis, electrodialysis (ED), adsorptive media, and anion exchange. Activated alumina (AA) is a selective ion exchange material that is often the method of choice for fluoride removal.

AA is a granular filter media that is primarily aluminum oxide (Al_2O_3). It has been activated by heat and washing with sodium hydroxide (NaOH; caustic soda). Fluoride ions in water are attracted to the surfaces of the AA. This is a pH-sensitive process in which the optimum is a pH of 5.5. Fluoride can be removed to less than 0.5 mg/L at this pH.

The capacity of the AA filters varies depending on the concentration of fluoride in the feedwater. As the capacity of the filter is reached, the media can be regenerated by washing with a caustic solution. The wash waste can be recycled, and the AA media can be used again to remove fluoride from the water.

Arsenic (III and V) is also removed from water by AA. If present in the water, arsenic will compete with fluoride for adsorption on AA. This is not often a problem because arsenic is usually present in very low concentrations when compared with fluoride. Therefore, it does not materially affect the capacity of the AA for fluoride removal. A modification of the AA treatment process can be used in the rare case in which arsenic is present in high concentration. Arsenic held on the AA requires higher concentrations of caustic for regeneration than does fluoride.

A typical treatment system consists of two down-flow pressure vessels in series. Fluoride removal is initially accomplished in the first vessel. As more water is treated, the fluoride-containing area of the AA in the vessels migrates down to the second vessel. The first vessel is then removed from service and regenerated. The second vessel continues to process water containing fluoride. When the first vessel regeneration is completed, it is returned to service in the lag position. In this way, continuous treatment is accomplished.

ADSORPTIVE MEDIA

Fluoride, arsenic, uranium, and other anions are often removed using adsorptive media. Specific materials used can be AA, modified AA, iron sorbents, and other proprietary adsorptive materials. The process operates similarly to the AA process described previously. The pH or other operating details may be different from the optimized AA process for fluoride removal. The system manufacturer should be contacted for operating specifics. Extensive pilot-testing should be conducted to verify the optimal operating conditions.

Source: Aqua Ohio Struthers WTP

Figure 10-5 Gravimetric fluoride feeder hopper

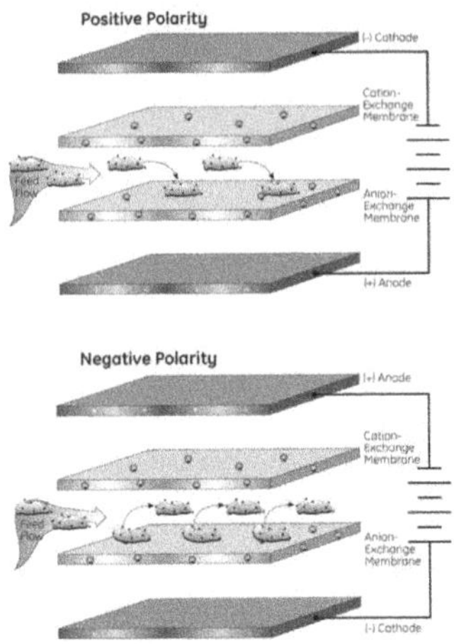

Figure 10-6 Schematic of electrodialysis

ED DEMINERALIZATION

Dissolved charged ions migrate through semipermeable membranes because of an attraction to electrically charged electrodes. ED can remove most dissolved ions from water and is an approved process to treat nitrate, nitrite, barium, selenium, and total dissolved solids (TDS).

Typical treatment systems require pretreatment with a scale inhibitor, chlorine residual of about 0.5 mg/L, and cartridge filter (10–20 µm). Sometimes air-stripping precedes the dialysis modules to remove any hydrogen sulfide.

The ED process relies on the fact that most dissolved ions in water are either positively or negatively charged. Selective semipermeable membranes allow either anions or cations to pass through while blocking those with the opposite charge. This results in a separation of the ions from the body of the water flowing through the system and concentrating the ions in the waste stream.

Electrodialysis reversal (EDR) is similar to ED, but the polarity of the electrodes is reversed regularly. This releases the ions from the electrodes and into the waste stream. This practice reduces scaling and increases the life of the membranes. The following are summaries of operating conditions and advantages of EDR/ED. See Figure 10-6.

Typical operating conditions of ED/EDR include

- turbidity less than 2 ntu;
- iron less than 0.3 mg/L (Fe^{+2});
- manganese less than 0.1 mg/L (Mn^{+2});
- hydrogen sulfide less than 1 mg/L;
- Silt Density Index (SDI) less than 15;
- free chlorine continuous 0.5 mg/L (tolerate up to 20 mg/L for a short time);

- total organic carbon less than 15 mg/L;
- pH 2–11;
- TDS up to 8,000 mg/L;
- some advantages of ED/EDR;
- minimal fouling or scaling;
- low pressure;
- long membrane life;
- tolerate nonionic species like silica; and
- low chemical usage for pretreatment.

PRECOAT FILTRATION (DIATOMACEOUS EARTH)

Diatomaceous earth (DE) is the skeletal remains of microscopic diatoms or algae. This material is mostly silica and inert, so it is suitable for filtering water. The process is approved for drinking water treatment.

The process uses DE to strain particulates from water and normally does not use coagulant chemicals. A thin layer of DE coats filter leaves or septum that is a plastic or wire mesh covering a hollow collection channel. The DE coating (precoat) is formed by a recirculating slurry of DE. There are two main types of DE filter systems: pressure and vacuum. Both have advantages and disadvantages.

Untreated water is applied to the precoat filter (usually about $^1/_8$ in. thick) with a small amount of DE added (called bodyfeed). Particulates in the water are removed on the filter surface, and this causes pressure to build (head loss). When the terminal head loss is reached, the filter is taken out of service for cleaning. High-pressure water spray on the filter cake is used to remove the accumulated material. This waste material is discarded, and the process is repeated, starting with the filter precoat with fresh DE.

DE filtration is used for source water with a maximum turbidity of 10 ntu. The filter loading rate can be between 0.5 and 2.0 gpm/ft^2. DE filtration is very effective for removing *Giardia* and *Cryptosporidium*. Dissolved substances and color are not effectively removed by DE filtration.

CORROSION AND SCALING CONTROL

The control or minimization of corrosion in water pipes and distribution systems is an important operational task. Corrosion is the deterioration of a pipe or fixture caused by its contact with water. The more corrosive the water is, the more likely that deterioration will take place. When pipes and fixtures corrode, they put materials into the water that may be harmful to the public (lead, cadmium), stain fixtures (iron, zinc, copper [Cu]), reduce the carrying capacity of mains and destroy hot water heaters, or produce sediments that harbor bacterial slimes, which in turn can cause tastes and odors and additional corrosion.

Corrosion occurs when all components of an electrochemical cell are present. These components include an anode, a cathode, a connection between the anode and cathode that allows electron transport, and an electrolyte solution that will conduct ions between the anode and cathode. If any one of these components is missing, corrosion will not take place. During the corrosion process, metal ions dissolve at the site of the anode. The electrons released in this process migrate to the cathode. Positive ions, in turn, migrate to the anode. This migration is the system's attempt to maintain an electrically neutral solution. The rate of migration affects the amount of metal that dissolves and is influenced by several factors. Knowledge and control of these factors, which are described in the following section, can help the operator keep corrosion to a minimum.

CORROSION FACTORS

The common factors that can affect corrosion are described in the following paragraphs.

- DO: In general, as the concentration of DO increases, the rate of corrosion increases.
- TDS: Water with a high TDS concentration conducts the flow of electrons. Pure water is not a good conductor. As the TDS of water increases, the rate of corrosion increases.
- Alkalinity and pH: Alkalinity and pH affect the rate of chemical reactions in water. A decrease in the corrosion rate is seen as the pH and alkalinity increase.
- Temperature: An increase in water temperature can increase the rate of corrosion.
- Flow velocity: Flow rates in pipes can positively or negatively affect corrosion rates. Where water is corrosive, a higher flow rate can bring DO to the surface of the pipe more readily and therefore increase the corrosion rate. However, when the water has been stabilized with corrosion control chemicals (lime, phosphates), a high velocity can decrease the corrosion rate by allowing for faster deposition of protective films.
- Type of metal: Metals are not usually found in pure form in nature. Rather, they exist as ores or other impure compounds. Corrosion of metals can be thought of as nature's way of bringing the metal back to a natural state. To add to the problem, certain metals naturally corrode more easily than others. If metals of high activity are combined with metals of low activity, the rate of corrosion between the two is increased. For example, when a copper service connection is coupled directly to an iron main, corrosion will take place; copper is much less active than iron, and iron corrodes more easily. Table 10-1 presents the galvanic series; in this series, each metal is anodic to, and may be corroded by, any metal below it. Table 10-1 illustrates that placing mild steel next to wrought iron would produce a slower rate of corrosion than would placing mild steel next to copper.

Table 10-1 The galvanic series for metals used in water systems

Corroded End (Anode)	Corrosion Potential
Magnesium	Most active ↑
Magnesium alloys	
Zinc	
Aluminum	
Cadmium	
Mild steel	
Wrought iron	
Cast iron	
Lead–tin solders	
Lead	
Tin	
Brass	
Copper	
Stainless steel	
Protected end (cathode)	↓ Least active

Source: AWWA 2013.

The technique of cathodic protection uses this principle. When two dissimilar metals are brought together in a water solution, the most active metal becomes the anode and will corrode. Water operators often put magnesium (Mg) alloy caps on iron pipe bolts. The caps become the sacrificial anode and are corroded. The bolts are thus protected at the expense of the caps.

LEAD AND COPPER RULE

Much of the lead and copper found in drinking water is leached from household plumbing that corrodes the water. The problem of toxic metals leaching from pipes, which occurs in drinking water, has been addressed by the USEPA in the Lead and Copper Rule (LCR). The health effects of lead and copper in water are well documented, and water suppliers are required to minimize the rates of corrosion of these metals. The LCR provides for periodic sampling from customer taps for the analysis of these two metals. Water that has been allowed to sit still in the pipes for at least 6 hours is sampled, and the values for lead and copper are listed in order of magnitude. The rule requires that 90% of lead levels be less than 15 μg/L, and 90% of the copper levels be less than 1.3 mg/L. If suppliers cannot meet these levels, treatment techniques that provide for corrosion control are required. Refer to Chapter 1 for further discussion of the LCR.

ION EXCHANGE

Groundwater is known to commonly contain high amounts of minerals and other natural substances that are present in the subterranean rocks and soil. Frequently high levels of hardness (calcium and magnesium) and dissolved radioactive ions are encountered and must be removed. Although precipitative softening (lime softening) can be used to effectively remove hardness, ion exchange is often used instead for smaller flows from many groundwater systems. Ion exchange softening is discussed in Chapter 9.

Radioactive ions, usually encountered in groundwater, are generally highly charged and therefore are ideal candidates for ion exchange removal processes. Regulated radionuclides susceptible to ion exchange removal include Radium 226 and 228, beta and photon emitters, which are usually present as cations (positively charged). The ion exchange process used to achieve their removal is similar to the ion exchange softening process.

Anion Ion Exchange

Some radionuclides are present in water as anions. Special test methods must be used to identify the radionuclide and determine the predominant species that is present. The most common radionuclide anion is uranium. Uranium is often present in water as the $UO_2(CO_3)_3^{4-}$ ion; depending on pH, the $UO_2(CO_3)_2^{2-}$ ion may also be present. Therefore, an anion exchange process can be very effective for its removal. The main differences between cation and anion exchange processes are that an anion exchange resin is used, and although sodium chloride is often the regenerant, it is the chloride (Cl^-) that is being exchanged. The disposal of concentrated regenerant for uranium is also a concern, as it is with other radionuclides. Refer to Chapter 9 for a discussion of ion exchange to understand the concepts of bed volume (BV), exhaustion, and regeneration.

A strong anion exchange resin is used for uranium removal. In many cases, the resin beds are able to effectively remove uranium for more than 200,000 BV. After the bed is exhausted, the bed resin can be disposed (care must be exercised not to exceed the radioactive limits for disposal). If the bed is regenerated instead, the run length is limited (often to 10,000–50,000 BV) so that the used regenerant meets disposal limits.

The concentrations of other anions (like nitrate, nitrite, or organic tannins) may also be removed by anion exchange. Depending on the conditions and amounts, these may compete with uranium and reduce the effective BV throughput. It may, however, be advantageous to plan on removing other anions in conjunction with uranium ion exchange treatment.

BIBLIOGRAPHY

AWWA. 2013 (4th ed.). *Water Distribution Operator Training Handbook.* Denver: AWWA.

AWWA. 2011. M58, *Internal Corrosion in Water Distribution Systems.* Denver: AWWA.

AWWA. 2011 (6th ed.). *Water Quality and Treatment* (J.K. Edzwald, editor). New York: McGraw-Hill.

AWWA. 2010 (4th ed.). *Principles and Practices of Water Supply Operations—Water Treatment.* Denver: AWWA.

AWWA. 2005. M53, *Microfiltration and Ultrafiltration Membranes for Drinking Water.* Denver: AWWA.

AWWA. 2004. M4, *Water Fluoridation Principles and Practices.* Denver: AWWA.

AWWA. 1988. M30, *Precoat Filtration.* Denver: AWWA.

Chowdhury, Z.K., R.S. Summers, G.P. Westerhoff, B.J. Leto, K.O. Nowack, C.J. Corwin, and L.B. Passantino, editors. 2012. *Activated Carbon: Solutions for Improving Water Quality.* Denver: AWWA.

Pizzi, N.G. 1995. *Hoover's Water Supply and Treatment.* Dubuque, Iowa: Kendall-Hunt Publishers.

USEPA (US Environmental Protection Agency). 2024. "TENORM: Drinking Water Treatment Residuals." www.epa.gov/radiation/tenorm-drinking-water-treatment-residuals (accessed September 2, 2025).

Chapter 11

Membrane Systems

Membrane system technologies have found increased use in the water supply community. Most contaminants found in water supplies can be removed using membrane technology. As the unit cost for installing membranes decreases and drinking water regulations become more stringent, more water utilities adopt the use of membranes.

The most common membrane processes in use today are reverse osmosis (RO), nanofiltration (NF), ultrafiltration (UF), and microfiltration (MF). RO is primarily used to treat brackish water or seawater, whereas the other processes are used to soften water or to remove disinfection byproduct precursors, turbidity, and pathogens.

Membranes work by providing a barrier against the passage of suspended, colloidal, or dissolved species found in source water. In general, the cost of membrane treatment increases as the size of the species being removed decreases.

This chapter focuses on NF and MF membranes for surface water treatment plants (WTPs) as explained by the Partnership for Safe Water. As discussed in Chapter 7, filtration is the critical unit process for removal of microbial pathogens. Although conventional filtration uses a combination of physical and chemical removal mechanisms, membrane filtration itself is solely a physical removal barrier through poor size exclusion.

Typically, MF membranes consist of one or more modules, and many individual fibers have a nominal filter pore size of 0.1 μm, whereas pathogens such as *Giardia* and *Cryptosporidium* are generally considered to be 1–2 orders of magnitude larger. Because of this, properly functioning membranes can be considered an extremely effective physical barrier capable of consistently providing a very high quality of treated water. However, membranes have a limited life span because their physical integrity deteriorates over time.

Unlike granular media filters or turbidity and particle counts, membranes require that additional data be gathered to gauge performance. A higher degree of sensitivity is required to identify individual fiber leakage that can allow the passage of particulate matter and present potential public health risk. Because of these principles, a simple self-assessment process for membrane treatment facilities requires the evaluation of additional data as compared with conventional surface WTPs.

Readers should refer to *Self-Assessment for Water Treatment Plant Optimization: Partnership for Safe Water* (Linder and Martin 2015), which can be obtained from AWWA. It is an excellent guide for the operations of membranes.

ULTRAFILTRATION

UF generally uses membranes with a pore size of about 0.002–0.1 nm. The molecular weight cutoff is in the 10,000–100,000 Dalton range, and the operating pressure is 30–100 psi. It therefore can remove colloids and other large molecular substances such as organic compounds. It does not remove hardness or salt but has the advantage of using less energy than the other systems described previously.

The principal advantage of UF is its ability to remove *Giardia* and *Cryptosporidium*, as well as viruses that may be present in surface water supplies.

MICROFILTRATION

The MF process generally operates with an effective pore size of about 0.03–10 nm. The systems have molecular weight cutoff of greater than 100,000 daltons and a feedwater operating pressure of about 15–60 psi. It can be used as a pretreatment process for other membrane units or to remove very fine particles in industrial applications, such as the electronics industry. Unlike the high-pressure RO and NF units, this lower-pressure system (including UF) must be backwashed frequently, which poses operational constraints on operators.

This technology is finding use in the direct filtration process because it can remove small particles that inhibit disinfection. In fact, an uncompromised system (one in which all membrane elements are intact) can achieve absolute removal of *Giardia* and *Cryptosporidium* organisms. Unlike UF, MF does not remove all viruses present in the surface water supply.

Some existing conventional WTPs, such as the Kenosha, Wis., facility, have opted to increase plant capacity by placing MF units in parallel with the conventional train. Kenosha's 40-mgd facility was increased to 60 mgd (approximately).

Membrane Integrity Testing

Low-pressure membranes (MF and UF) are commonly used for drinking water treatment. The membrane surface, although quite uniform, may have imperfections or may develop gaps during operation that could allow the passage of some contaminants through the system. Concern has been amplified where the membrane is the primary barrier in the treatment process.

Integrity testing of membranes is a necessity to provide assurance that the system is operating as intended to provide a positive barrier to the passage of pathogens. This testing can be direct, indirect, and can be conducted while the system is in operation and on-line, or during maintenance operation and off-line. Some systems use several testing methods to provide added confidence to the integrity of the system.

Direct test methods include the pressure decay test, diffusive air flow, the bubble point test, the acoustic sensor test, the liquid–liquid porosimeter test, the air hold test, and the binary gas integrity test. Indirect tests include particle counting, particle monitoring, turbidity monitoring, and surrogate challenge tests. Several of these test methods have been adapted to online monitoring.

Most of the integrity tests now available are for the low-pressure membrane systems. Similar methods for high pressure (NF and RO) are under development. Indirect methods are being used routinely. Research is being conducted to establish integrity tests that are most useful for these systems.

New or refined test methods are under development. Many of the integrity tests are specific to the membrane type or even the manufacturer. Each system has its own requirements, and tests are used that are the most suitable to the situation. This is an area that will continue to evolve and improve. The goal is to have a continuous online monitoring system to ensure membrane integrity.

POSTTREATMENT

Treatment of membrane product water consists of different unit operations for RO and NF systems. Primarily, the treatment schemes may include aeration, pH adjustment, disinfection, and filtration. There may also be a need for alkalinity recovery and elimination of hydrogen sulfide, if present. Note that there is little or no demand for chlorine (Cl) after membrane treatment, so operators will need to dose at the level they need for the distribution system. Alkalinity recovery, which is needed because of the destruction of alkalinity in the acid step, is beneficial to the system in terms of corrosion control and Lead and Copper Rule compliance. Product water stability should be controlled to reduce corrosion.

BIBLIOGRAPHY

AWWA. 2024. M61, *Desalination of Seawater*. Denver: AWWA.

AWWA. 2022. M46, *Reverse Osmosis and Nanofiltration*. Denver: AWWA.

AWWA. 2019. M53, *Microfiltration and Ultrafiltration Membranes for Drinking Water*. Denver: AWWA.

AWWA. 2011 (6th ed.). *Water Quality and Treatment* (J.K. Edzwald, editor). New York: McGraw-Hill.

AWWA. 2010 (4th ed.). *Principles and Practices of Water Supply Operations—Water Treatment.* Denver: AWWA.

Linder, K., and B. Martin. 2015. *Self-Assessment for Water Treatment Plant Optimization: Partnership for Safe Water.* Denver: AWWA.

Liu, C. 2012. *Integrity Testing for Low-Pressure Membranes.* Denver: AWWA.

Wolfe, T.A. 2001. Personal communication.

Chapter 12

Testing and Laboratory Procedures

Operators, refer to AWWA M12, *Simplified Procedures for Water Examination*.

REASONS FOR TESTING

Water treatment plant (WTP) operators take samples every day. Water in the plant and the distribution system is sampled frequently for two reasons: for compliance testing to ensure the water is safe for human consumption, and to measure the efficiency of the treatment processes. Some samples, such as those for disinfectant residuals and turbidity, are tested onsite and continuously; others, such as those for disinfection byproducts (DBPs) or metals, may be sent periodically to a laboratory for analysis. The size and resources of the individual utility will affect the choices that the operator makes regarding sampling.

SAMPLING

Reliable test results depend on proper sampling procedures and techniques. Samples are supposed to accurately represent the quality of the water being sampled. Consequently, consideration must be given to the way the sample is collected, the volume taken, any storage requirements for the sample, and sampling locations and frequencies. Usually, the regulating authority will disseminate standard procedures for sampling and analysis, and the utility will be responsible for maintaining laboratory certification.

Sample Collection

Improper sampling is a common cause of error in water quality analysis. Because the results of an analysis can only show what is actually in the sample, the sample must have the same content as the water from which it was taken. Water laboratory analysts refer to this as a *representative sample*, and it is defined as being one of two types: grab sample or composite sample.

Grab samples are single-volume samples collected at one time from one place. An operator taking a 100 mL sample from a distribution system for bacteriological analysis is taking a grab sample. Similarly, an operator taking a few milliliters of settled water for turbidity analysis is taking a grab sample. As such, these samples represent the quality of water in those locations only at the time that the sample was obtained. For this reason, grab samples are best used when the quality of the water is not expected to change significantly or where changes take place slowly. A grab sample might not give an accurate representation of the quality (over time) of a flowing river that is subject to intermittent pollution from upstream wastewater plants or chance rainfall. The frequency of grab sampling is determined by studying the history of changes in water quality so that adequate representation of the quality can be made.

A composite sample consists of multiple grab samples taken from the same location at different times and mixed together to make one sample. Time-composite samples are made up of equal-volume grab samples taken at different times. An example of a time-composite sample is 100-mL samples of backwash water taken at 1-min intervals for the length of the wash and mixed together to measure total solids. Flow-proportional composite samples consist of different volumes of a sample taken from the same location at different times; the volume collected each time is dependent on the flow rate of the water at the sample location. When composite samples of source water in a WTP are needed and the source water flow rate varies over the day, a proportional composite sample is sometimes taken because it more accurately represents the water quality over time.

Composite samples allow the laboratory analyst to determine the average concentration of a constituent over time without having to perform the analysis every time a sample is taken. Automatic samplers can be used to take samples at night, which enables the chemist who comes to work in the morning to perform the analysis. There are drawbacks to composite samples. The high and low points, or ranges, of a contaminant cannot be known because they are averaged. Also, some constituents, such as pH, dissolved gases, chlorine (Cl) residual, and temperature, begin to change immediately after the sample is taken and therefore should not be composited. Bacteriological samples, such as those taken for coliform analysis, must always be taken as grab samples because the number of bacteria in the samples begins to change immediately after sampling.

Sample Volume

The volume of sample will vary according to the testing procedures used and is often set by regulations. For example, coliform compliance sampling requires a 100 mL grab sample. The laboratory analyst will provide instructions for sample volume.

Sample Storage

Glass and carbonate (CO_3^{-2}) plastic sample containers are most often used in water utilities, but glass is avoided for cases in which there is a greater chance of breakage, such as in shipping. Also, glass is not used when sampling for fluoride because fluoride will adhere to the glass and not provide accurate results. Plastic can also be problematic because certain organic chemicals can permeate the plastic, and therefore, the organic chemicals will not be detected in the subsequent analysis. Some types of plastics may actually release organic chemicals into the sample and thus invalidate the results.

Samples for total trihalomethane (THM) and haloacetic acid (HAA) analyses are taken in glass vials, usually colored brown to filter out sunlight, and shipped in coolers. Often, sample containers have preservatives, quenching agents, or pH-adjustment chemicals in them to facilitate sample preservation and/or storage, which extend the shelf life of the sample, allowing more time for analysis. This, in turn, allows the analyst to collect all the necessary samples and then perform a single analysis, greatly reducing errors that may otherwise occur during calibration. It is also more efficient and less costly to perform one calibration for multiple samples than it is to calibrate for each individual sample.

Regulated parameters have specific directions for sampling, preservation, and storage before testing. These requirements are listed in the regulatory agency compliance regulations. Some agencies have adopted the US Environmental Protection Agency (USEPA) requirements, and some have separate, additional requirements. Many, including USEPA, have referred to *Standard Methods for the Examination of Water and Wastewater* (2022). Consult the regulatory agency for requirements in the utility's location.

Sample Location and Frequency

Selecting representative sampling points and determining sampling frequency are important steps in attaining meaningful water quality data. Samples are usually collected from three areas within the treatment system: the raw or source water, the plant, and the distribution system. Because sources and treatment schemes differ from system to system, no guidelines are given in this handbook. Local regulatory agencies usually provide a minimum frequency for sampling, and most operators take samples with a greater frequency than is required.

In-plant sample locations should be chosen to allow for the measurement of the performance of each unit process. Operators should take care not to sample immediately after chemical addition but rather at a point when proper mixing has been achieved. Sampling of combined filter effluent turbidity should take place before posttreatment chemicals are added to avoid measuring added turbidity, such as from lime or phosphates.

Distribution system samples are intended to measure the quality of the water that is delivered to customers at various points in the system. In general, samples should not be taken from customers' taps. The exception is for lead (Pb) and copper (Cu) testing and for water quality complaint sampling. The inlet and outlet of storage tanks are good sampling sites because they give information about the effects of storage. The direction of flow should be noted when sampling, i.e., whether the tank is filling or emptying at the time of sampling. Bacteriological sampling and the associated chlorine residual testing are regulated. Frequency of sampling is based on population served.

QUALITY ASSURANCE/QUALITY CONTROL

Standard Methods (2022) defines quality assurance (QA) as "a definitive plan for laboratory operation that specifies the measures used to produce data of known precision and bias." It defines quality control (QC) as a "set of measures within a sample analysis methodology to assure that the process is in control."

A QA program of a laboratory consists of a QA manual, written procedures, work instructions, and records. For example, an organizational chart that lists the training, capabilities, and responsibilities of each lab analyst would be part of a QA program. QC is also part of the overall QA program.

In a QC program, each analyst demonstrates his or her capability to obtain acceptable results for an analysis. This is demonstrated by analyzing a set of "blind" standards. Analysts are given samples that contain a specific amount of a chemical. The analyst runs the samples, and the results are reported to an agency that knows the correct levels of chemicals in the samples. The agency then rates the laboratory and the analyst on their performance. This type of testing is commonly called a *performance evaluation* (PE). PEs are typically run twice a year and determine if a laboratory is "certified" to run the analysis.

Another type of continuous laboratory evaluation is performed through the use of "known" standards. Known standards are samples that contain a predetermined amount of a chemical. These standards, along with each set of samples, are run by an analyst. This allows an analyst to determine how well a particular set of analyses has been performed in each analytical trial. Power fluctuations or other daily hazards may affect laboratory equipment, and the evaluation of known samples will let an analyst determine if a particular run is satisfactory. Sometimes these known samples are referred to as *standards*. For example, known turbidity standards are run before each set of filter turbidity grab samples is run, thus allowing an operator to determine if the instrument is properly measuring turbidity.

A final set of samples is also used to determine data quality. These samples are referred to as *spiked samples*, *laboratory control samples*, or *laboratory-fortified matrix samples*. Spiked samples are made in the laboratory by adding a known quantity of the chemical being measured to the collected samples. The analyst splits one sample into two aliquots and then adds a precise amount of the chemical being analyzed to one of the aliquots. The two aliquots are run and the results compared. The sample to which the chemical was spiked should have a level of chemical equal to the known addition plus the results of the unaltered sample. This type of sample is important because it allows the analyst to determine if some other chemistry in the sample is interfering with how the instrument measures the contaminant. Regulations and reference to *Standard Methods* (2022) determine the frequency of this type of QA sampling. Different analyses require different numbers of QA samples; this will affect the number of samples an operator may be required to take for any given parameter.

Another way to evaluate laboratory performance is by reviewing historical analytical performance. This is known as *QC*. The results of known samples and spiked samples are graphed and reviewed for trends over time. This is important because the efficiency of analytical equipment can decrease over time, or an analyst can improve his or her performance with experience. For example, a decrease in the amount of a known sample result over time can indicate that a probe or detector needs to be replaced.

The importance of laboratory QA/QC to overall plant operations cannot be overstressed. It is critical that test results be reliable because they are a measure of how a plant is performing and a measure of the safety of the water the plant is producing. Performance must be measured accurately, and the accuracy of analysis depends on sample quality and analytical quality. All laboratories should strictly adhere to the laboratory practices described in this section and be willing to produce QA/QC information as requested.

Common Water Utility Tests

The following is a brief description of some of the more common parameters that WTP operators frequently measure, including their significance. When applicable, cross-references to other chapters are given.

Alkalinity

The buffering capacity of the water is measured by its alkalinity. This test can determine the concentration of carbonate (CO_3^{-2}), bicarbonate (HCO_3^{-}), and hydroxide (OH^{-}) alkalinity. These measurements are useful to determine the corrosive nature of the water in combination with other factors and to optimize the lime–soda ash softening process.

The test involves careful titration of a measured quantity of water with a standard 0.02*N* sulfuric acid solution to pH end points of 8.3 and 4.5 (these are indicated by color change indicators or by using a pH meter). The carbonate, bicarbonate, and hydroxide alkalinity are then calculated from the results.

Carbon Dioxide

Water that contains high concentrations of free carbon dioxide (CO_2) can cause the consumption of lime when using this method of softening. Also, carbon dioxide is a factor affecting corrosion.

The test usually used is to titrate a measured water sample with a standard solution of sodium hydroxide (NaOH). The end point is signified by a change in color for phenolphthalein or a pH of 8.3.

Chlorine (Free or Total)

For WTPs that use chlorine for disinfection or oxidation, this is one of the most important tests performed by operators. After any addition of chlorine, a measurement should be taken routinely. This test will verify the correct dosage and reveal changes that may affect plant performance and the safety of the water supply. Many plants use both free and total (or combined) forms of chlorine in their processes.

The test most often used is the N,N-diethyl-1,4-phenylenediamine sulfate (DPD) color test. The DPD (either for free or total chlorine testing) is added to a water sample, and the intensity of the color indicates the amount of chlorine present in the sample. Most plants use a digital read-out colorimeter to give an accurate result. Some color comparison portable test devices are used, as well; however, they are not always accurate. Another test method is amperometric titration. This test is usually performed in the laboratory. A special meter is used to determine the end points when titrating a measured sample with a standard phenylarsine oxide solution (Figure 12-1). This method is very accurate and is capable of determining many chlorine species that may be present. There are several other chlorine residual test methods (*Standard Methods* 2022) that can be considered.

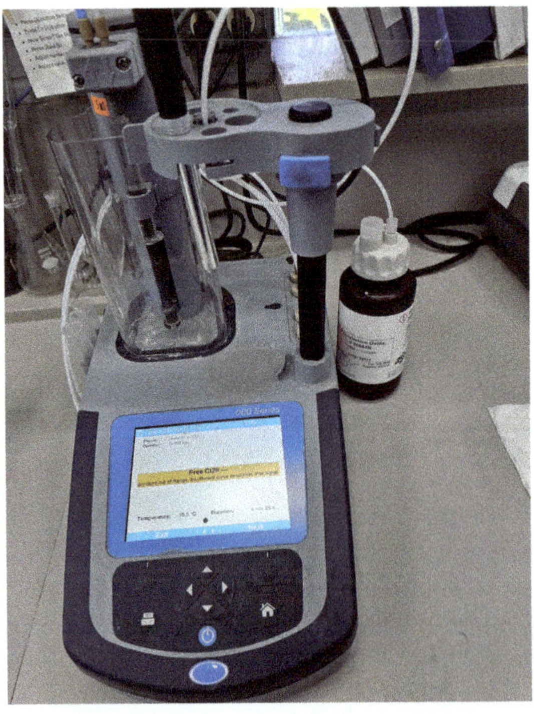

Source: Aqua Ohio—Struthers Division

Figure 12-1 Laboratory titrator for chlorine residual

Both the DPD color test and amperometric titration are used in online instruments for continuous chlorine monitoring. As with any instrument, these devices must be calibrated and checked frequently to ensure accuracy.

Chlorine demand can be determined using the residual test method. This measurement can be used to predict residual chlorine over a specified time. Jar testing apparatus may be used for this test. A sample is taken, and the chlorine residual is measured immediately. After a specified time, the residual is measured again, and the difference is the demand. Care is needed to duplicate the conditions (light, temperature, holding time) that are of interest.

Chlorine Dioxide

See Chapter 8 for a discussion of chlorine dioxide disinfection, generation, and operation. Also, Chapters 9 and 10 include some information regarding ion exchange for iron (Fe) and manganese removal and greensand filter operation. Chlorine dioxide residual is limited to 0.8 mg/L (USEPA maximum residual disinfectant level), and a byproduct, chlorite, is also regulated with a maximum contaminant level (MCL) of 1.0 mg/L.

The test methods for chlorine dioxide are similar to those for chlorine. There are DPD and amperometric test methods, but some of the test conditions have been modified to yield chlorine dioxide–specific results. Also, there is an ion chromatography method that requires trained technicians in a certified laboratory.

Coliform

Coliforms (see Chapter 8) are a group of bacteria that produces gas bubbles in lactose or lauryl tryptose broth at 35.5°C within 24–48 hours. They are considered indicator organisms, and their presence may indicate the presence of other, more harmful bacteria and organisms. Because total coliforms are easier to analyze in the average water utility laboratory than the actual disease-causing micro-organisms, total coliform testing is used in place of more tedious, more expensive testing for these other organisms.

Most laboratories use one of two methods to test for coliforms. One method is the membrane filter technique, in which water samples are passed through a 0.45-µm filter. The filter paper is fine enough to trap bacterial particles as the water passes through. The filter paper is then placed onto growth media (such as M-endo) and incubated. Any bacterial colonies that are present will grow in size, be visible to the naked eye, and can be counted after a period of time (usually 24 hours).

Another method is the MMO–MUG (minimal medium) technique. This method allows for inoculation of water sample bottles with powder. The substance in the bottle will feed any total coliform that may be present and produce a color change during incubation.

Samples for coliform testing are always collected in sterile bottles and in quantities sufficient for testing. Most bacteriological samples require a minimum of 100 mL (approximately 4 oz) for analysis. Samples should be analyzed the same day as they are collected but can be refrigerated for 8 hours before analysis. Coliform samples are taken in the plant at various stages to test for process efficiency; they are also taken in the distribution system for regulatory compliance. The number of samples that must be taken in the distribution system is a function of the population served. Many WTPs also take coliform samples of the raw or source water.

Much of the bacteriological labware can be purchased in presterilized condition and then sent to waste. However, any bacteriological equipment that is to be reused must be sterilized in steam using an autoclave, as shown in Figure 12-2.

Conductivity

This test (more correctly named *specific conductance*) measures the ability of water to conduct electricity. This is an indirect measure of the ions or minerals in the water. It is sometimes used to estimate the total dissolved solids (TDS) content because of its ease of measurement and the availability of inexpensive mobile field instruments.

A conductivity meter is connected to an electrode, and this is immersed in a water sample. An instrument setting to match the sample temperature is often used to adjust for extremes. The conductivity reading is in microsiemens/centimeter (μs/cm).

Source: Conneaut, Ohio, Water Department

Figure 12-2 Autoclave used for sterilization of reusable glassware for the coliform tests

Cryptosporidium

Cryptosporidium is a parasite regulated under the Long Term 2 Enhanced Surface Water Treatment Rule (LT2ESWTR). It is a regulated pathogen, and its measurement is necessary to ensure adequate treatment depending on the occurrence in the untreated water supply.

The method involves filtration, specialized separation, and identification methods. This test can only be performed by laboratories approved for this method. Plant operations personnel will use the test results to determine compliance with the LT2ESWTR.

DBPs

Several organic compounds are created when chlorine is used in water treatment. Two groups of these byproducts are regulated: THMs and HAAs. Compounds in both groups have been classified as probable carcinogens.

The test methods for both groups involve procedures that must be performed by trained analysts in approved laboratories. Compliance testing is, therefore, not usually performed by plant operating personnel. Online instruments are available for continuous monitoring, but the instruments should be checked periodically by the same approved laboratory procedures. Operator-performed screening tests can be used to indicate levels for plant monitoring, but these methods are not approved for compliance testing.

Dissolved Oxygen

The amount of dissolved oxygen (DO) in the source water may be an indicator of the condition of the lake or reservoir being used. Lack of oxygen (O_2) may be a predictor of water quality problems caused by anoxic conditions. The DO concentration in the treated water may be a factor contributing to the rate of corrosion.

The DO test involves the use of a dedicated meter and specific sensor probe. Also, there are color methods in which the intensity of the color from the addition of the indicator is a measure of the concentration present in the water.

Fluoride

Fluoride reduces tooth decay. Fluoride may be naturally occurring or added during water treatment as directed by local health authorities. The optimal amount of fluoride needed to provide health benefits depends on several factors. There is a maximum MCL of 4.0 mg/L for fluoride in drinking water.

There are both instrumental and colorimetric tests for fluoride in water. Operations personnel usually use the color methods. The color indicator is added to the water, and the color intensity measures the amount of fluoride present.

Giardia lamblia

Giardia lamblia is a parasite found in untreated water supplies, and it is pathogenic. It is regulated by a treatment technique in which disinfectant concentration and contact time are prescribed to ensure adequate inactivation. Separate requirements are specified for chlorine, chloramine, ozone (O_3), and ultraviolet.

The method involves filtration, specialized separation, and identification methods. This test can only be performed by laboratories approved for this method. Plant operations personnel will use the test results to determine compliance with the Surface Water Treatment Rule.

Hardness

Hardness is mostly attributed to the amount of calcium and magnesium in the water. Softening treatment plants must carefully monitor hardness as a process control. Most other water systems regularly monitor for this parameter as a basic measure of water quality.

There are several methods for hardness testing. Although hardness can be determined by separately analyzing the calcium and magnesium levels, most water systems use a color titration method for total hardness results. A measured sample is titrated with a standard solution (ethylenediaminetetraacetic acid) to a color change end point.

Inorganics (Heavy Metals)

Some metals are toxic in high amounts (such as cadmium or chromium), and others are not (such as iron or manganese). There are MCL regulations for many heavy metals and secondary limits for those that can cause aesthetic concerns. Most regulated metals are included in the primary drinking water regulations, and some have special rules such as lead and copper.

The test methods for metals are tremendously varied. Instrumental and color methods predominate. Only approved methods can be used for compliance testing. Operations personnel may elect to use unapproved screening methods for process monitoring. The more precise methods require trained analysts and must be performed by approved laboratories.

Iron and Manganese

These metals are often grouped together because they are similar in their role in water quality and often occur together. Both cause staining of fixtures and are not generally toxic at normally encountered concentrations. Treatment processes have been developed to remove one or both of these substances. Secondary standards (non-enforceable) are established for both.

There are several test methods for both of these metals. Many treatment plants use a color method for process control. These methods are typical when an indicator is used, and the intensity of the color is measured to determine the amount present.

Lead and Copper

These metals are grouped together because of the USEPA regulatory rule requirements. Both substances are toxic at elevated levels, and thus, they are regulated. Generally, these contaminants are somewhat unique in that they are not often found in source water but are the result of dissolution from piping components (see the Lead and Copper Rule description in Chapter 1). Therefore, testing is mostly performed in the distribution system rather than the treatment plant.

The test methods vary from instrumental laboratory procedures to simple field-test color methods. The laboratory methods require trained analysts and are not usually conducted by operations personnel. Screening tests using colorimetric methods are often used by plant operators for process control.

Microbiological Organisms

Many microbiological organisms are found in untreated water. Bacteria, viruses, protozoan, algae, and plankton are examples of the multitude of possibilities. Some of these are pathogenic; others can produce toxic substances. The array of consequences is very broad. Some are regulated in drinking water. The coliform group of bacteria is used to indicate the possibility of contamination. Another useful bacteria indicator is the heterotrophic plate count. This is a general bacteria population measurement that may indicate the presence of other pathogens.

Test methods are as varied as the organisms. Many involve growing colonies on selective nutrients and counting the number from a measured sample. Others may be examined by microscope and identified by experts. Some chemical byproducts are analyzed by instrumental methods capable of detecting minute amounts. Generally, training analysts must perform most of these procedures. The exceptions are the coliform tests that have been developed using color changes to both detect and enumerate these bacteria.

Nitrate, Nitrite, and Ammonia

These three inorganic nitrogen compounds are often encountered in water supplies. Contamination from agricultural activities is often the source in surface water, but another significant source is wastewater discharges. Nitrate and nitrite are regulated contaminants with enforceable MCLs. Ammonia is not regulated in drinking water but is a compound that may encourage the growth of nitrifying bacteria. Ammonia is often associated with the chloramination process.

Several test methods are available for these substances. Operational control testing is usually conducted using colorimetric methods. Special precautions are needed when testing for free ammonia in the presence of chloramines to ensure accurate results.

Ortho- and Polyphosphate

Phosphate can be naturally occurring in surface water where it may be the result of urban runoff or waste discharges. Orthophosphate and polyphosphate are often used as corrosion inhibitors in water supplies. These substances are not considered toxic, but the amount is carefully monitored to ensure optimal effectiveness.

There are instrumental methods for phosphate that require approved laboratories and trained analysts. Most operational control testing uses colorimetric tests. Selecting the correct test may require knowledge of the type of phosphate used for corrosion control.

pH

This is one of the most common tests performed in water treatment and drinking water monitoring. The pH value (Chapters 5, 9, and 10) is an indicator of the acidity or alkalinity of the water. There is not an MCL for pH in drinking water. However, there is a secondary standard range of 6.5–8.5.

pH uses a scale from 0 to 14, with the midpoint of 7 being neutral—i.e., the acidity and alkalinity are balanced. At less than 7, the acidity of the water predominates; at greater than 7, the alkalinity of the water predominates. With each unit increase or decrease, the concentration or intensity changes tenfold.

For example, for the pH to change from 5 to 6, the acidity must decrease by a factor of 10. The pH of water is significant because it affects the efficiency of chlorination, coagulation, softening, and corrosion control. Also, pH testing can provide early warning of unit process failure. For example, the addition of alum to the rapid-mix stage should produce a predictable drop in pH. If it does not, a malfunction of coagulant feed could be indicated.

Samples for pH should be collected in glass or plastic containers and analyzed as quickly as possible. Samples should not be agitated because dissolved carbon dioxide could be liberated, which will change the pH.

A pH meter is used for the test in combination with a suitable probe. The meter must be periodically calibrated using a known standard. Figure 12-3 shows a laboratory-scale pH meter being calibrated with a pH 4 stock standard solution. The pH value for a water sample may change while standing because of a change in temperature or exposure to air. Therefore, measurements are usually taken immediately after sampling.

Radiologic Substances

Contamination from radiological substances is more common in groundwater than surface water. Several of these substances have MCL regulations. There are numerous possible radioactive isotopes that may be encountered; therefore, screening water supplies for general radioactivity (gross alpha and gross beta activity) may be prudent before using specific substance testing.

Testing for radioactive substances often requires specialized equipment and procedures. Approved laboratories are required for compliance testing, and analysts with specific training are needed. Test results for radioactivity are sometimes expressed as pCi/L (picocuries/L) but, depending on the substance, µg/L or mrem/y may be used.

Solids—Settleable, Dissolved, Suspended

The solids contained in a water supply are all the substances that are not H_2O. These can be inorganic, organic, volatile, nonvolatile, suspended, settleable,

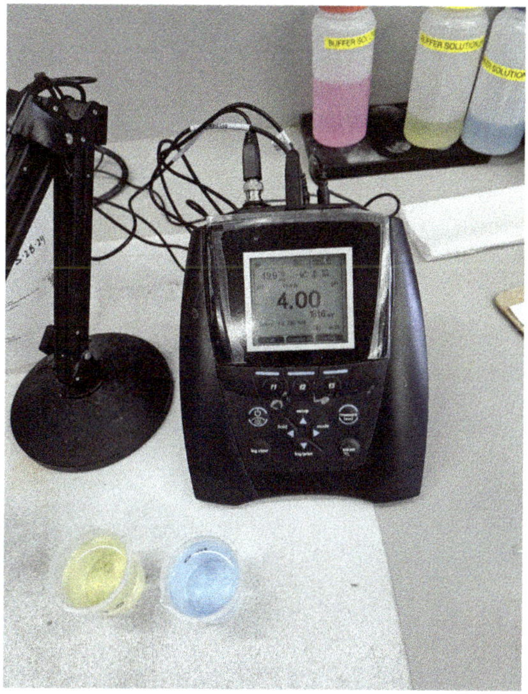

Source: Aqua Ohio–Struthers Division

Figure 12-3 Lab pH meter being calibrated with a pH 4.0 standard solution

or dissolved. Water is most often sampled and analyzed for turbidity as a measure of clarity with a standard turbidimeter (Figure 12-4). However, some water plant side-streams (e.g., supernatant from wash water return basins) are so turbid that they are best analyzed for suspended or dissolved solids. There are no MCLs for solids content, but there is a secondary standard for TDS of 500 mg/L.

The test method involves taking a measured sample and heating it to remove the water and then weighing the remaining residue (solids). The temperature used for heating can define the volatility of the residue. Also, filtration before heating can be used to give a dissolved result. TDS is the most used solids measurement in drinking water. TDS is a gross measure of the inorganic content of the water (because organic substances are usually a minor part of the total).

Synthetic Organic Chemicals

This is a large group of organic compounds that is regulated and has MCL standards. Many of these are pesticides and herbicides that mostly are not currently manufactured in the United States. Most of these compounds are probable carcinogens and, therefore, have MCL goal limits of zero. These

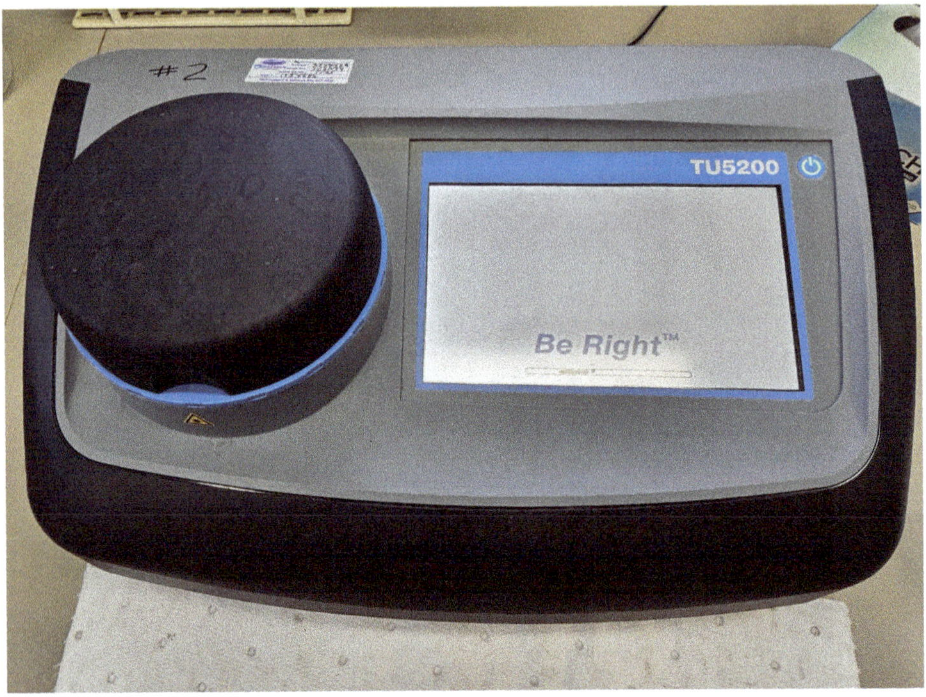

Source: Aqua Ohio–Struthers Division

Figure 12-4 Laboratory turbidimeter

must be tested according to the requirements of the Primary Drinking Water Regulations.

The test methods for this group of compounds involve complex chemistry instrumental procedures. Only approved laboratories with trained analysts can perform these tests. A few of these substances have intricate rapid test methods that are not approved for regulatory compliance. These methods may be suitable for screening surveys or other occurrence evaluations.

Taste and Odor

Consumers often react to taste and odor as their only way to evaluate the quality of the drinking water. A disagreeable response may result in a poor perception, so it is imperative that the water utility seek to provide water that is pleasing to its customers. There is no regulatory MCL for taste or odor, but there is a secondary standard MCL for threshold odor number (TON) of 3.

There are several tests for taste and odor. One is the flavor profile analysis. This test involves a trained panel to routinely evaluate the water. Several utilities use this method with good results. However, most utilities find this method to be labor-intensive and instead use the older TON method. This method enlists a panel to smell water samples of various dilutions. The consensus of the dilution for which an odor was detected is the TON for the sample. Although this test can be a bit subjective, it can also be useful to detect odor problems and to assess the effectiveness of treatment.

Temperature

This is probably the test most frequently performed on drinking water. There is no MCL for temperature. Differences in temperature can indicate probable water quality problems. Also, water supply changes are often linked to temperature. On large bodies of water such as Lake Erie, a sudden downward swing in temperature can indicate a dead zone incident, where colder water from the lake bottom has made its way to the intake.

A thermometer is used and the water must be tested at the sample location. The result is usually expressed in Celsius because this scale is used for many other test procedures.

Total Organic Carbon

This parameter is a nonspecific measure of the organic content of the water. There is no MCL for total organic carbon (TOC), but the Disinfectants and Disinfection Byproducts Rule uses TOC measurements to determine if precursor removal is needed to comply with the rule. Enhanced coagulation or other means may be required to reduce the TOC.

The test uses a TOC instrument either in the laboratory or online. Calibration is required by analyzing known standards and adjusting the instrument to provide an accurate result. Specific training is necessary to calibrate and use the instrument to provide the best results.

Total Suspended Solids

Operationally, this is a simple test that determines the amount of solids left in the water after it has moved through a dewatering process such as sedimentation, filtration, sludge thickening, etc. For water treatment, this test is not used as much as turbidity testing except in high-solids situations, like those found in precipitative softening plants, for example. In those cases, it is convenient to arrange the sample piping in a way that allows the operator to more easily take grab samples. Figure 12-5 depicts a good example.

Volatile Organic Chemicals

Several volatile organic chemicals are regulated and have MCLs. Although these compounds are volatile, it is possible that water under pressure may contain these substances, and when the pressure is released, the compounds could be inhaled.

Test methods for this class of compounds involve careful sample collection (zero headspace). The samples are purged with an inert gas, and the vapor is analyzed by gas chromatograph (instrument). This test must be performed in an approved laboratory by trained analysts. Operational personnel may assess the results and compare them with regulatory standards.

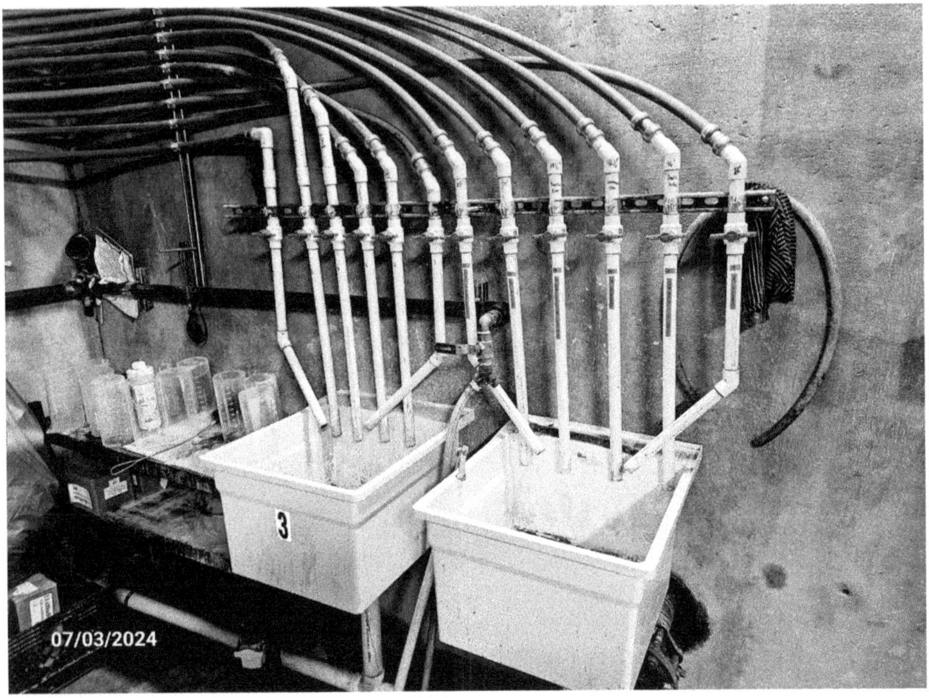

Source: Aqua Ohio—Struthers Division

Figure 12-5 Sampling station for total suspended solids samples from solids-contact units at WTP

Continuous Read Instrumentation

Taken as a whole, supervisory control and data acquisition (SCADA)-controlled WTPs depend heavily on the data that is being provided on a real-time continual basis. These instruments must be kept in top working order and must be calibrated. A real problem with this process is that there are many alarms set to keep the operator informed. It is not uncommon to hear multiple alarms sounding. The constant din tends to bring on a "fatigue" of sorts, during which the operator may subconsciously become so acclimated to it that it gets ignored. Humans tend to prioritize these alarms and choose which ones to attend to and which to set "on the back burner" for later attention, which often gets forgotten. Good training is important here. Placing the instrumentation readouts in one convenient location (Figure 12-6) encourages the operator to check all of the parameters when responding to the alarms.

Jar Testing

Jar testing is perhaps the least understood test but the most useful process tool available to operators. A few simple ideas and techniques need to be mastered. When the plant staff becomes efficient in this process control

Source: Aqua Ohio—Struthers Division

Figure 12-6 Wall-mounted instrumentation gathering various water quality parameters that are being sent to SCADA

strategy, they can rapidly respond to treatment upsets. This is especially important with the more stringent turbidity requirements of the Interim Enhanced Surface Water Treatment Rule.

Jar testing can be used for many tasks, such as screening new coagulants and polymers, coupon testing for corrosion control, biological spiking experiments, and more. The jar testing described in this handbook focuses on coagulant dosage control and oxidant demand. In cases in which turbidity breakthrough in a WTP has occurred, an examination of the records often indicates that the staff did not respond adequately to a sudden change in source water turbidity or a sudden source water demand for oxidant. In these cases, the staff simply did not know what dosage of chemicals to use to counteract the changes because they could not perform jar testing.

Jar testing requires a working knowledge of stock solutions. The following text explains how to prepare stocks for alum, alum–polymer blends, ferric chloride ($FeCl_3$), chlorine, and potassium permanganate ($KMnO_4$). It also shows how to perform jar tests with these stocks to determine coagulant dosage and chlorine or permanganate demand. The jar test apparatus is the device most commonly used. It consists of a series of six 2-L square jars

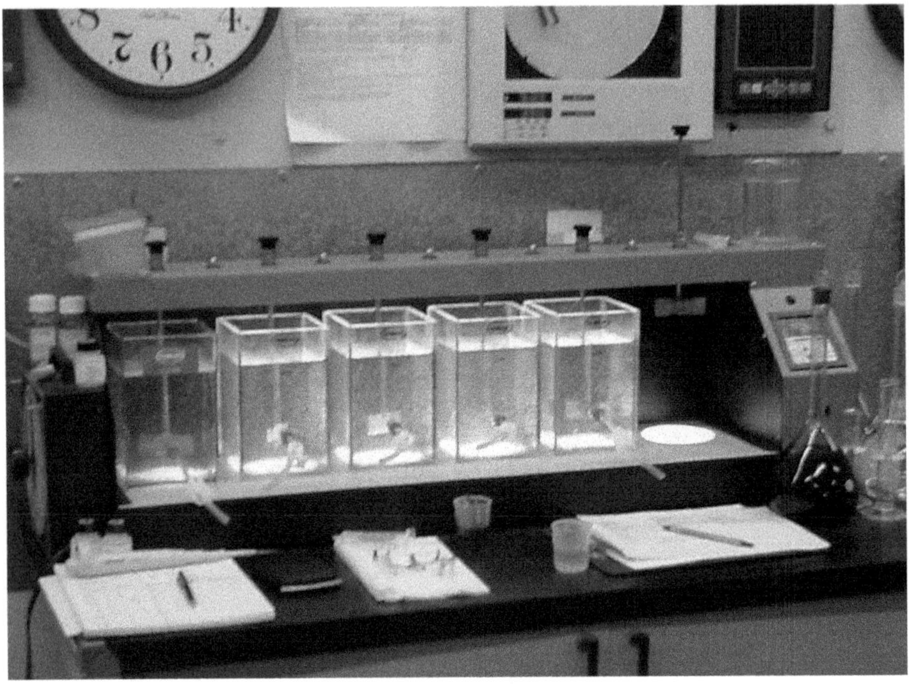

Source: Cleveland, Ohio, Division of Water

Figure 12-7 Jar test equipment used for determining coagulant dose

with stirrers. A common shaft operates the stirring mechanism, ensuring that each jar is mixed identically. Operators should follow the procedures outlined in the manufacturer's recommendations for operation.

Using Stock Solutions

Jar testing equipment is used to determine proper coagulant dosage and for process trial-testing. Square jars provide better mixing characteristics than round jars. Figure 12-7 shows one such setup. Stock solutions are used in these tests, and their preparation requires great care.

Alum Stock

Alum dose is usually expressed in milligrams per liter as dry basis. If liquid alum is used, the product is shipped to the plant as approximately 48.8% alum (8.3% Al_2O_3). Commercial alum is approximately 642-mg/mL dry basis. To make a working stock solution for jar testing, pipette 15.6 mL into a 1-L volumetric flask and dilute to the 1-L mark with distilled water. 642 mg/mL × 15.6 mL is approximately 10,000 mg, so this stock is 10,000 mg/L, or 10-mg/mL dry-basis alum. Every milliliter of this stock placed into a 2-L beaker for jar testing would produce a dosage of 5-mg/L alum dry basis.

Alum–Polymer Blend Stock

Stocks made from alum-polymer blends should be prepared so that operators can express dosage as "parts per million"—or ppm—which means that 1 gal of coagulant is added to each million gallons of source water. This expression negates the requirement to know the exact proportions of alum and polymer in the coagulant, which may be proprietary information not shared by the vendor.

Stock solutions should be made fresh each week at minimum. Refer to Chapter 4 and AWWA Manual M37, Operational Control of Coagulation and Filtration Processes, for greater details and discussion.

Liquid alum–polymer blend

Alum weighs about 10.9 lb/gal (it may vary), so 18 gal × 10.9 lb/gal = 196.2 lb/mil gal

196.2 lb/mil gal/8.34 = 23.5 mg/L of liquid alum–polymer blend

Because roughly 45% of this is dry alum, 45% × 23.5 = 10.6-mg/L dry-basis alum

To make a working stock, place 10 mL of the commercial product into a 1-L volumetric flask and fill to the mark with distilled water. This is a 10-parts-per-thousand stock. Each milliliter of this stock placed into a 2-L jar is a dosage of 5 ppm.

Ferric Chloride Stock

Ferric chloride dosage is usually expressed as a milligrams-per-liter product. The material generally comes to the plant as a liquid with a stated percentage that can vary according to manufacturer. Most NSF-certified ferric chloride products are limited to a dosage of 250 mg/L; however, most plants use far less.

To make a working stock solution of ferric chloride from the dry powder, dissolve 2.93 g into 1,000 mL of water and mix thoroughly. The resulting stock is 1,000 mg/L or 1 mg/mL. One milliliter of this stock added to a 2-L jar is a dosage of 0.5-mg/L dry ferric chloride. To make stock solutions from the liquid product, it is best to consult the supplier.

When it is desirable to compare alum and ferric salt coagulants for enhanced coagulation comparisons, make a working stock solution from the liquid product (assuming 38% ferric chloride, $FeCl_3$, specific gravity 1.4), transfer 10.27 mL to a 1-L flask, and fill to the mark with laboratory-grade water. This solution is 10-mg/mL ferric chloride. One milliliter of this solution put into a 2-L Gator jar will be a dosage of 5 mg/L. The calculation is 38% is 38 g/100 g, and when multiplied by the specific gravity, it yields a concentration of 0.532 g/mL. At 0.532 g/mL × 10.27 mL, a solution is produced that is 5.5-mg/L ferric chloride, which is equivalent to a dosage of 10-mg/L alum. Because the enhanced coagulation requirements are written in 10-mg/L increments of alum, this will allow the operator to make easy comparisons.

Chlorine Stock

Gaseous chlorine (stored as compressed liquid) is used for disinfection and for chemical oxidation. The gas is fed through chlorinators equipped with site tubes graduated in pounds per day or in pounds per million gallons. Dosage for chlorine is calculated using the following formula:

$$\text{Lb Fed/mil gal/8.34} = \text{mg/L}$$

The impurities in the water will exert a demand for chlorine, so that the residual amount in the water will be less than the amount fed. Other factors, such as temperature and detention time, will also affect the demand for chlorine. For this reason, operators should know how to test for chlorine demand.

To make a working chlorine stock using commercial bleach, make an approximate 2,000-mg/L chlorine stock by adding 40 mL of bleach to a 1-L volumetric flask and fill to the mark with deionized (DI) water. Verify the strength of this stock by adding 1 mL of it to 1 L of DI water. This should test as 2-mg/L free chlorine residual because DI water has no chlorine demand to it. Whatever it tests at, record and use that number when spiking jars for chlorine.

To perform a chlorine demand test, 2 L of source water should be placed into a square jar and stirred at a slow speed. While stirring, add 2 mL of the stock to the square jar and time the process. After 15 min, obtain a sample from the jar for chlorine analysis. The dosage minus the residual is the demand. For example, if the stock solution prepared as described tested as 1.8 mg/L, then 4 mL of that stock placed into 2 L of source water would produce a dosage of 3.6 mg/L. If, after 15 min, the residual chlorine of the jar is 2.0 mg/L, then the demand is a ratio of 3.6 to 2.0, or 1.6 mg/L.

Note that when performing a simple chlorine demand test, no other oxidants should be added to the jar. Also, commercial bleach has a pH greater than 10 and may increase the pH of the water in the jar.

Potassium Permanganate Stock

Potassium permanganate is used as an oxidant for iron and manganese control, some organic precursor control, and for taste-and-odor control. Permanganate will enhance the coagulation process as it begins to oxidize some of the constituents in the source water. It is thought that the continuous use of this chemical will further reduce the need for chlorine at the front end of the plant. Dosage of permanganate is calculated like any other dry chemical: Pounds fed divided by million gallons, in turn divided by 8.34, will provide a dosage as milligrams per liter. Potassium permanganate has the unique quality of a color change mechanism, which signals to the operator its effectiveness and end point. A pink color is associated with water freshly dosed with permanganate. This color changes to straw or yellow at the end point. Operators can use this knowledge as an aid for dosage and demand predictions. Permanganate reactions are dependent largely on pH.

Like chlorine, there is a demand for permanganate. Therefore, operators can use jar testing techniques to calculate this demand.

Make a 1-mg/L stock solution each week by weighing 1 g potassium permanganate into a 1-L volumetric flask and diluting it to 1 L. This is a 1-mg/mL working stock. Obtain a source water sample taken before the chemical addition point for potassium and pour it up to the 2-L mark in a Gator jar at the jar test station. Add 1 mL of stock permanganate and agitate for 5 min at slow speed on the stirrer. This is a 0.5-mg/L dose. Add the appropriate amount (current operational amount) of stock alum solution and continue to agitate for another 20 min. Observe color change during that time.

Sample the settled water from the beaker and run a free chlorine residual test on it. Multiply the result by 0.89; this is the milligrams per liter of permanganate left in the sample. Subtract this from 0.5 mg/L (dosage). The result is the amount that should be fed. For water of very little demand, adjustments may be necessary.

Coagulant Jar Testing

Preparation for Jar Testing

NOTE: The following are generic procedures for jar testing. Because each WTP is different, the procedures will have to be customized (scaled down) by the operators to fit the particular system.

The purpose of jar testing is to simulate full-scale plant operations. Successful jar testing depends on the ability of the operator to simulate the conditions of the WTP as closely as possible. Important considerations are those of proper dosage, proper chemical addition sequence, proper mixing times, and maintenance of operating temperatures. Also important is the need for good sampling and analysis techniques and application of the data produced to the full-scale operation.

Operators should determine the times and sequences of the jar testing procedure using knowledge of their particular plant. These sequences will provide for test results of higher quality if they are accurate. AWWA M37 is an excellent resource for helping the operator determine times and sequences that should be used for testing. The main reason that jar testing fails to produce suitable results is that operators make the mistake of assuming that detention times for settling that are found at plant scale must be duplicated at the bench scale. In reality, bench-scale settling times should be adjusted, or scaled down, and AWWA M37 shows how to do that. Most importantly, the sedimentation basin overflow rate must be known because it determines the length of time that floc should be allowed to settle in the jars before sampling. Because overflow rates are a function of plant flow, jar test results can take on new meaning each time an operator makes a change in pump rate.

Starting with fresh stocks, fill weigh boats or pipettes with the proper amount of chemicals and place them next to each jar before proceeding. If using a programmable jar test apparatus, it should be preset for the times that are being used. Manual machines will have to be controlled by an operator at each step change.

Fill each Gator jar with 2 L of fresh source water just before testing. The source water should be obtained from a point in the process before any chemical addition. Alternatively, if permanganate is being used and the testing being performed is for turbidity removal results only, source water with the permanganate already in it can be used.

Jar Test Procedure

A generic procedure for jar testing for coagulant dosage control is provided in this section. Plant operators should alter these procedures and times to suit their particular needs. Consult AWWA M37 for a more detailed jar testing procedure.

Begin the testing by setting the mixers to moderate speed. Operators should have calculated detention times of each sequence of their plant.

Add the permanganate (if desired) and allow mixing for a time equal to the travel time (baffling factor adjusted) to the rapid mix. If applicable, add chlorine stock.

Add coagulant and set mixers to full speed for the duration of the detention time that has been chosen. Be sure to add the coagulant to all jars simultaneously.

Turn down the speed of the mixers to simulate flocculation energy for the duration chosen.

After flocculation, turn the mixers to the lowest speed possible to simulate flow through the settling basin.

After 5 min, samples can be taken for analysis. The most common analysis is the test for turbidity.

Tips: For colder water temperatures, samples of source water just before jar testing should be taken so that the operator will work with the coolest water possible. If the jar testing is being performed for TOC removals, be sure to wipe the mixer shafts with a clean cloth to remove any oils that may have seeped from the bearings of the mixer. If more than two jars are being tested, it is advisable that two operators perform the testing. This allows for the simultaneous application of chemicals to the jars. Use a clean syringe to take samples from the tops of the settled jars rather than using the attached hose. An excellent reference for jar test procedures is the Partnership for Safe Water manual, available from AWWA.

BIBLIOGRAPHY

AWWA. 2011 (3rd ed.). M37, *Operational Control of Coagulation and Filtration Processes*. Denver: AWWA.

AWWA. 2010 (4th ed.). *Principles and Practices of Water Supply Operations—Water Quality*. Denver: AWWA.

Owen, C.A. 2001. Personal communication.

Pizzi, N.G. 2007. *Pretreatment Pocket Field Guide*. Denver: AWWA.

Standard Methods for the Examination of Water and Wastewater. 2023 (24th ed.). Washington, D.C.: APHA, AWWA, and WEF.

USEPA (US Environmental Protection Agency). 1989. "Drinking Water—National Primary Drinking Water Regulations—Total Coliforms (Including Fecal Coliforms and *E. coli*)." 40 CFR 141. *Federal Register*, 54:124:27544.

Water Supply Operations: Jar Testing. DVD. 2010. Denver: AWWA.

Chapter 13

Instrumentation and Control Equipment

The design of a functional instrumentation and process control system for a water treatment plant (WTP) begins by establishing a basic control philosophy. The designer needs to know the period of time that the plant will operate (continuous versus partial days), the degree of automation and process control desired, the type of signal to be transmitted, and whether the plant will be operated using local or remote control. Some knowledge of the types of alarms to be used and other telemetry requirements and computerization desired will also be necessary. A WTP, with all its subprocesses, is a specialized system. Knowledge of these subprocesses and their relation to each other leads the design engineer to take a comprehensive approach.

This chapter acquaints the operator with that approach, so that he or she can assist and comment appropriately during the design phase. Operator involvement during the design phase can be quite useful. An operator should realize that the design of the facility, which takes a few months, will affect operations for as long as the plant is in service. An operator's involvement at this critical stage may save effort and difficulty later.

PROCESS CONTROL

The main purposes of process control are to produce a quality product, be alerted to process or equipment malfunction, and provide data acquisition that allows for plant evaluation and reporting. Data acquisition is also used to make decisions on process changes, which are necessary from time to time. Plant size, difficulty of treatment, and the capability of the operating staff affect the types of controls specified. Availability of qualified maintenance and electronic technicians, whether in-house or outsourced, will also affect the choices that are made.

Control Methods

Controls provided in the plant can be manual, semiautomatic, automatic, and supervisory. It is a good idea to choose the least complex methodology available that meets the control objectives. This will differ from plant to plant and allows for more onsite operator involvement, which is important in process control. There is a tendency for some design engineers to provide control systems that take the operator away from the process. This is done with good intentions—the engineer desires to ease the operator's tasks—but can lead to problems.

Consider this: the process engineer who designs the components of the WTP may never have operated one or may never have had to cope with the issues in the supervision of one, such as shift work, sick-time call-out procedures, employee issues that arise, power outages, and all of the day-to-day stumbling blocks that get in the way of well-intentioned and well-designed process control configurations. It is a virtual certainty that the WTP that is being designed will at some point have to be operated under manual mode and under stressful conditions. If an operator wants to be able to handle the plant at such times, it is a good idea for the operator to participate in the beginning of the conceptual design phase. The designer probably does not have operating experience and therefore cannot entirely meet the operators' needs.

An example is the complete automation of a filter backwash process. It has become standard design to allow for filter wash to take place based on predetermined criteria (head loss, turbidity breakthrough, etc.) with no operator involvement. All the valve turnings, sequences, pumping, and other necessary functions will proceed without flaw, freeing the operator to complete other tasks. The designer views this as positive. However, in that instance, the operator may not view the backwash for "hot spots," hose down the walls and piping during the wash, sample the floc as it travels over the weirs, view the "rate of rise," or notice any of the many problems that can occur during the wash. That filter wash may come at an unpredictable time or at a time that is forced because of any number of other process issues. An operator does not learn about backwash, or any process, without witnessing it and controlling it.

Supervisory Control and Data Acquisition

Control systems found in many WTPs are referred to as supervisory control and data acquisition (SCADA) systems (Figure 13-1). The importance of SCADA is enhanced in part because of the regulatory compliance framework of the Interim Enhanced Surface Water Treatment Rule (IESWTR) and other similar requirements. Gathering, archiving, and analyzing huge amounts of raw data, which is required for compliance, require a more sophisticated data system such as SCADA.

SCADA systems are powerful and diverse, but it is important to choose these systems carefully based on the true benefit to the WTP. Poorly designed systems can cause operational problems, and so-called state-of-the-art systems can be obsolete in just a few years, making it difficult to find replacement parts.

Photo courtesy of Cleveland Division of Water

Figure 13-1 Operator checking SCADA control console

There is real concern when the SCADA technician (integrator) does not work in concert with the operator, or when the operator does not know exactly what to ask for in process data acquisition capabilities. What often results is the operator being controlled by the SCADA system. The staff often has no control over the data being displayed, recorded, and archived. The operator may not have control over system set points once the integrator has left the scene, and that can create a long and miserable night for a hapless operator.

For example, Figure 13.2 shows a WTP data screen for filter #1 depicting a time frame of roughly 50 min one morning during a filter run. The original graphic is in color, so each line is easily distinguished. For black and white pictures, some "callouts" have been added for reference. There are five total lines; four have been labeled with a callout. The fifth line, mostly straight across the top (at 0.75), is filter level, which is not important for right now.

Engineers and integrators created this format, and it is clever in its depiction of five filter-related data points that are being monitored and shown. The five data point labels are displayed in the lower left-hand corner as FLOW MGD, TURB NTU, LOH, EFFLUENT VALVE POSITION, and FILTER LEVEL. The operator can temporarily see one item at a time by touching the item name and holding it. When an item is touched, the y-axis scale even changes to more closely fit the "spread" of data values. We can tell by looking at the upper left-hand corner and by the high and low values shown that the last person to use the screen wanted to see turbidity for filter #1. If you look at the five columns of values just to the right of the turbidity label,

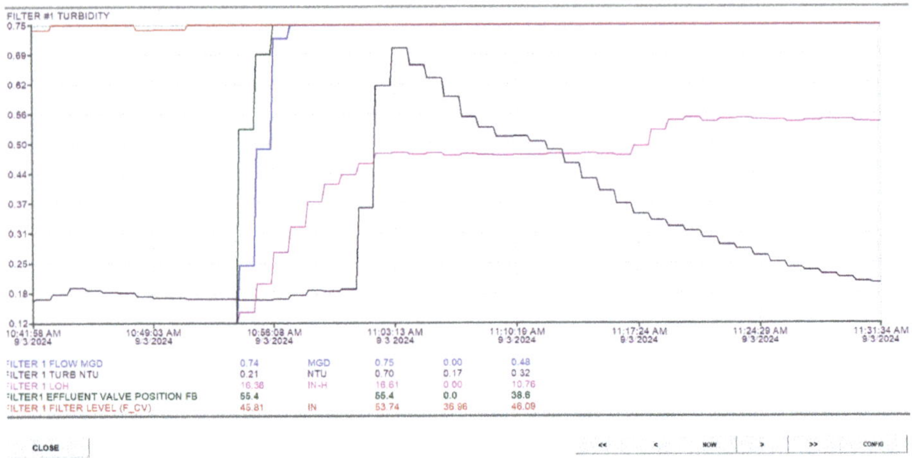

Source: Nick Pizzi

Figure 13-2 Filter turbidity output on SCADA display

you'll see that the first two columns show us that currently the turbidity is 0.21 ntu; the high value was 0.70 ntu, and the low value was 0.17 ntu. The last column is an average for the time window that is displayed. The other four items will disappear as long as the operator is touching the label. That provides less distraction for the operator and is a nice addition. This is valuable information for an operator. It encourages the ones who want to improve their system performance, it teaches the new operator some of the basics, and it sparks curiosity in some who may seek to find meaningful trends in the data. But beyond that, there is little else the operator can do with this screen display. What other features would make things better for the operator? Let's discuss it from the operator's viewpoint. Let's talk to operators who have experience, know the filter regulations, and are committed to optimizing their WTP processes under Partnership for Safe Water guidelines. What might some of those operators say?

Turbidity

It seems that the filter was put on-line at about 10:55 a.m. Why was turbidity being recorded before the filter was in service? The regulation states that turbidity is to be recorded during the filter run, not when off-line or during backwash. The recorded data are required to be available to the regulator for three years. It would be better, and maybe easier, to explain what happened during a filter run a couple of years ago if the data were truly indicative of the water going to the clearwell. Perhaps the line could be changed to a dashed line or simply eliminated during those periods.

Also, when it comes to filter effluent turbidity, the use of the mathematical function "average" is meaningless and can even be misleading. A person looking at this tiny slice of time could walk away thinking that "on average, these people put out 0.32 ntu!" SCADA integrators are fond of averaging numbers. It is important that the operator explains that values like "95th percentile" and "MAX" are more meaningful when they pertain to turbidity and can lead to better operations.

The Other Values

It is obvious that when the operator chose the screen for turbidity, the resultant scale got set to a range that proved to be meaningless for flow rate and valve position. What is the benefit of that? Could the graph have a second y-axis that could display another value or two? True, the real-time number can be found in the columns at the bottom, so a graph isn't necessary. But if that's true, why was one created to begin with?

Some operators like numbers; others prefer pictures. Graphical displays like these help some operators view trends of the treatment process. Operators who have worked in a WTP for a long time realize that individual data points are fine and necessary. But a trend of the data, what happened an hour, or a shift, or a day, or a month ago, can tell you where the process is headed, and is really the most important thing for an operator to know as soon and as clearly as possible.

The truth about this scenario that we've been working with is that the shift operator cannot control the SCADA system's finer points. The integrator must be called in, for example, to retrieve the data and put it into a spreadsheet for data analysis. This makes complying with the Partnership for Safe Water requirements a frustrating exercise in futility. It seems ludicrous, but in many cases, the utility does not truly "own" its own data: a private contractor does.

Reliance on SCADA

A real consideration is the complexity of the treatment process. In general, WTPs are not as complex as industrial chemical plants, and the need for high-level SCADA may be questionable. However, this concept must be balanced with the size of some plants; very large facilities generate large amounts of data, which may be a justification for a more complex system.

Manual control of process components such as mixers, flocculators, and some valves is desirable. It is important to consider the restart of this equipment after a power failure. Continuous-operating equipment, like a flocculator, should be automatically restarted if the power fails. Operator safety, however, is an overriding criterion that must be considered whenever automatic restart is chosen. Semiautomatic controls are specified for processes such as initiation of backwash or other batch processing. Automatic and supervisory controls, which lend themselves to rate-of-flow control, chlorination, and sludge handling, must be carefully chosen.

In many cases, it may be a better choice to require some functions to be set and restarted manually. This is, of course, a decision to be made by the operator of record, and it should be made carefully by that person after all due consideration of human resources and viability of the plant equipment.

The bottom line is this: if the WTP is a complex set of processes that take knowledge, experience, and skill to work properly, then it is best if each operator learns how to run the plant manually before the SCADA system fails and the operators lose control of the processes.

Control Strategies

The two basic types of automatic controls are discrete control and continuous control. Discrete control ties equipment status (on–off) to status changes or programmed events. An example of discrete control is the automatic start–stop sequence for high-service pumps based on clearwell level. Continuous control requires an analog measurement for its input and manipulates a final control as an output. An example of continuous control is the feeding of chemicals based on fluctuating source water flow. Chemical dosage is increased or decreased based on the flow rate signal.

Continuous control takes many forms, and they are discussed in the following sections.

Feedback Control

In the classic control-loop function of feedback control, a controlled variable is "fed back" to the controller for the purpose of detecting deviation from a set point. This system checks and rechecks actual system output against a desired value. In this way, it automatically corrects deviation, thus proving its value to the operator. The limitation of this system is that correction can only be made after deviation has occurred, and elements such as meters, transmitters, and transducers are needed.

Feed-Forward Control

In feed-forward control, a process is controlled without the use of feedback; the control action is initiated before deviation occurs. Many WTPs use this type of control for simple chemical feed. The feeder output is controlled in proportion to plant flow, and the dosage goal is set manually at a ratio station. This simple controlling system can be used where accuracy is not critical and when time lags in the system will not take on great importance.

Two-Position Control

The simplest form of automatic feedback control is the two-position feedback system. It is called *on–off control* but is not limited to on–off situations. An example of its use is found in on–off pumping that is based on reservoir level. It is used when close control is not required.

Floating Control

Floating control is a three-position control that is used where control action is reversible. It is a variation of the on–off control but has the added elements of a second "on" position and a dead band. An example is the control of valves where a sequence of opening–off–closing is needed. This application is valuable because of its ability to accept operating times, such as the amount of time necessary to open a valve. It offers a tighter amount of control, or narrower range, than on–off control and will usually cycle to a lesser degree.

Floating Proportional Control

Floating proportional control is the on–off control equivalent of a conventional analog controller, with automatic reset features included. The advantages it has over analog systems are the simplicity of the on–off control and the fact that it will not lose control action when signal is lost.

This is a great advantage to the operator because it can eliminate sudden process upset.

Proportional Control

Simply stated, a proportional controller uses a mode in which the output of the controller is directly proportional to the deviation from the set point. This type of system is used frequently for filter level control.

Proportional Plus Reset Control

The inherent limitation in proportional control is its inability to eliminate deviation from the set point over the control range. To overcome this, reset control is added. The reset function of this controller attempts to constantly change the controller output if there is deviation from the set point. When the process needs a wide proportional band for stable operation, this application is specified.

Proportional Plus Reset Plus Derivative Control

Proportional plus reset plus derivative control increases the response of the proportional plus reset control and is used only on slow-responding processes. These controllers are seldom used in WTPs.

Accuracy

The degree to which an actual value will conform to an indicated value is the accuracy of the system. Accuracy, expressed as a percentage, is usually defined or specified as a percentage of a range, a percentage of an actual value, or a percentage of a range of a measured variable.

Accuracy and precision (repeatability) are not the same. Figure 13-3 shows the difference between the two. The figure depicts two targets that have been used by two sharpshooters. Sharpshooter A has placed five shots near the bull's-eye; one shot has found its mark. The other four are close, close enough in fact to consider sharpshooter A to be accurate. Sharpshooter A was not able to repeat the performance—he was not precise—but his performance may have been acceptable for the application intended.

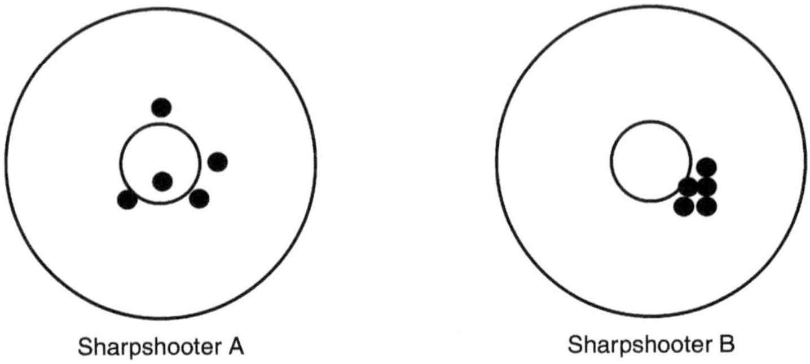

Figure 13-3 Sharp shooter diagram

Sharpshooter B was very precise in his pattern. He did not hit the target, but all of his shots were placed in a tight pattern. His performance was repeatable but unacceptably inaccurate. He needs to adjust his rifle sight to find success.

Real-Time Measurement

Process control is accomplished using measurement or detection devices. On-line instrumentation allows for continuous measurement of a process and therefore for immediate and accurate control of that process. Popular on-line measuring devices include pressure indicators, level indicators, on-line turbidimeters, chlorine (Cl) residual recorders, and flow indicators. Off-line instruments, such as laboratory devices, although not commonly used for controlling a process, are important to the operator. They help measure the effectiveness of a process for compliance purposes and serve as calibration devices for on-line instrumentation.

Level

Operators need a reliable and accurate method of measuring levels in the plant, whether those levels are in the clearwell, the day tank, or some other location. Common methods for measuring and transmitting levels are float-type level indicators, bubbler units, pressure-actuated level transmitters, and sonic-type indicators. Each has advantages and disadvantages, and operators have their personal likes and dislikes. The simplest device is the float-type sensor. However, it is not practical where freezing will occur, and it requires maintenance to provide a relatively friction-free motion for the cable assembly. Sonic devices, which work readily on a 4- to 20-mA system and require temperature correction, are popular.

Pressure

Gauges and transmitters for pressure sensing work in one of two ways. The first method uses a manometer, which balances the weight of a liquid, such as mercury, against the pressure to be measured. The second method

Photo courtesy of Aqua–OH Water Co

Figure 13-4 Complex pump arrangement requires multi-faceted flow and pressure monitoring with remote access capability.

measures the distortion of a metal caused by the pressure of water against that metal. Pressure is measured as absolute, gauge, or differential. Absolute pressure is referenced to an absolute vacuum, which is zero. Gauge pressure is measured with reference to the ambient atmospheric pressure, which is about 14.7 psi at sea level. Differential pressure shows readings that are the difference between two pressure readings.

Manometers use a U-shaped glass tube, which is etched with gradations of pressure. The tube is filled with mercury, which is 13.6 times as heavy as water. The pressure of the water pushes the mercury up the tube, and a reading is taken. Because of the cost and hazards associated with mercury, manometers are being replaced with other types of pressure sensors.

Most pressure-sensing devices used in WTPs rely on the metal-distortion principle. Differential pressure cells also use this method. Water pressure is introduced to one side of a metal diaphragm, and the distortion is measured and translated as pressure, or pressure differential.

Flow

Flow is the most expensive and expensive physical measurement taken in the plant. Cheap measuring systems often result in high head loss measurements and inaccuracies; more expensive devices can be quite inflexible. The flow rate of water in a treatment plant is measured using various types of meters, including loss-of-head meters, positive-displacement meters, magnetic flowmeters, Venturi meters, rotameters, flumes, and weirs. For large flows, Venturis and flumes or weirs are most often chosen. They are nonmechanical devices that need little maintenance. Very low flow rates, such as for chlorinators, are usually measured by rotameters.

Filter Head Loss

Measuring loss of head in a filter is important because many operators terminate the filter run based on this measurement (see Chapter 7). Most plants have a filter strategy in place that requires filter run termination based on a preset loss of head. However, usually operators will terminate based on unit filter run volume and filter effluent quality (nephelometric turbidity units [ntu], particle counts) because of public health considerations. Nevertheless, it is necessary to measure head loss in a filter. Head loss development that deviates from the norm or historical data can provide the operator with valuable information.

Particulate Contamination

Particulate contamination in water is measured to determine the efficiency of the treatment process and as proof of compliance with regulations. As discussed in Chapter 1, the Surface Water Treatment Rule and the IESWTR regulate the turbidity of the finished water. A goal of treatment is to reduce, as much as possible, the amount of source water particulate contamination before the water is delivered to the customer.

Turbidity

Turbidity is an optical property of water. When light is directed into a water sample, some light passes through the water, whereas some light is scattered. Some light is also absorbed by particles and dissolved substances in the water. Some turbidimeters and turbidity measurement methods provide correction for interferences from light-absorbing substances. Some do not. Water with high turbidity will scatter more light and appear cloudier than low-turbidity water.

Turbidity is measured in one of two ways: by measuring the light that is scattered or by measuring the light that passes through.

The first method, called the *nephelometric method*, is commonly used in WTPs because it is sensitive and precise in clearer water. The second method is outdated and not commonly used in WTPs.

Turbidity measurements are taken with bench-top, portable, and on-line instruments. For regulatory reporting purposes, on-line and bench-top units are used. On-line instruments are commonly used to monitor the effluent of individual filters, whereas bench-top instruments are often used to monitor raw, settled, and combined effluent, or “plant tap,” water. As previously mentioned, the IESWTR requires that all filters in surface WTPs be monitored with on-line instrumentation.

Bench-top turbidimeters are batch-sample analyzers designed for grab samples, although some instruments can be equipped with flow-through cells that measure continuously. Bench-top units use glass cuvettes that hold approximately 30 mL of sample. The cleanliness of the cuvettes is critical for obtaining meaningful measurements. If multiple samples are taken and measured at one sitting, *indexing* (the reading contribution from imperfections in the sample container, cuvette) must be used to minimize the deviation in readings brought about by the optical properties of the glass cuvette. New cuvettes should be tested for this deviation; any cuvettes that do not match should be discarded. Cells that deviate more than 0.01 ntu when measuring low-level turbidity should be rejected.

Cuvettes must be free of dirt, water spots, and fingerprints and should not be stored in an open container. Silicone oil is applied to the glass surface for cleaning and to minimize spots. Some operators use devices (sonicators) to degas water samples, i.e., remove air, which can contribute to false readings. Care must be taken when using these devices because they can cause dirt particles to break off the cell walls of the cuvette.

Instruments should be calibrated according to manufacturer's instructions, which are based on constructing a calibration curve that covers the range of expected measurements. Calibration is accomplished using primary and secondary standards. Primary standards are prepared from traceable raw materials and include formazin, approved by the US Environmental Protection Agency, and as required by state regulatory agencies, for this purpose. Quarterly use of primary standards is usually required. Secondary standards are those that a manufacturer has certified to give instrument calibration within a certain tolerance when the instrument has been calibrated with the primary standard. Secondary standards, which are used daily for bench-top instruments, are very stable.

On-line turbidimeters sample a sidestream that is drawn from the treatment process. The water being measured flows from the process to the instrument's analyzer chamber, and measurements are made continuously. On-line turbidimeters may be used at any point where a water supply can be tapped to provide representative, reliable sample flow and where the instrument itself can be accessed for routine maintenance and calibration.

Particle Count

Particle counting provides the WTP operator with the potential to enhance the monitoring and control of water treatment processes. Although there is debate about the significance of particle counts and removal efficiencies as they correlate to pathogen removal, most utility operators who use particle-counting devices believe that they enhance plant operation. Correlation between particle counts and turbidity in water varies depending on the cause of the particles or turbidity. Specifically, turbidity fluctuates more as a result of the smallest particles in water (less than 1 μm in size, such as colloids commonly found in treated water), and particle measurements fluctuate more because of larger particles such as coagulated solids that break through or off filters or form postfiltration via chemical precipitation.

The operator who effectively uses both on-line turbidimeters and particle counters will come to better understand the total particle characteristics of filtered water. For example, through the course of a filter run, smaller particles (as measured by turbidity) will continually decrease; near the end of the run, larger particles (as measured by particle counters) will begin to detach. This detection of elevated particles before turbidity breakthrough allows the operator to further optimize filter operations.

A goal for filter effluent production is set by choosing total particle count. For example, an operator might measure particle counts for particles larger than 2 µm and attempt to keep those counts less than 50/mL in the filter effluent. When the filter effluent's particle count is greater than 50/mL, the filter is taken out of service, and backwash variations are implemented in treatment chemistry or filter backwash procedures. It has been advocated that operators track and respond to increases in particles in specified size ranges, such as those that are 3–5 µm, as a possible surrogate for *Cryptosporidium*. Because particles of all sizes have been shown to be removed with similar methods, it is usually best practice for operators to track and respond to total particle counts (for particles smaller than 2 or 3 µm, depending on the lower detection limit of the counter). This gives the largest counts to work with and, for this reason, provides the greatest potential for noticing and responding to any elevated counts.

Particle counters operate by light-blocking, or by light-scattering, principles. Those commonly used in drinking water analyses may provide different counts because of their specific characteristics, e.g., calibration or make. There is little agreement currently as to calibration standards. For this reason, particle counts are not regulated, as is turbidity, but particle counters can be a very useful tool to compliment turbidimeters for use in treatment optimization.

Temperature

Temperature can affect process control systems, laboratory analysis, and equipment operation. The measurement of temperature is a component of regulatory compliance when calculating $C \times T$ (concentration times time) (see Chapter 8). Temperature of the backwash water must also be measured to ensure efficiency in the bed-expansion portion of the wash. Temperature is commonly measured using thermocouples, thermometers, and bimetallic units. Thermocouples operate on the principle that an electromotive force is generated when heat is applied to one of the junctions of two dissimilar metals. The electromotive force, which is proportional to the difference in temperature at the two junctions, is measured and displayed as temperature.

The thermometer, or thermal bulb, operates on the principle that the absolute pressure of a gas is proportional to the absolute temperature. Bimetallic units are constructed of two different metals bonded together. Each metal expands and bends at a different rate and thus responds differently to different temperatures. The amount of bend is measured and converted to temperature.

Chemical Measurements

pH

The term pH is defined as the negative logarithm of hydrogen (H) ion activity, and it covers a scale of values from 0 to 14. The measurement of pH requires an electrode, which is composed of a pH-sensitive device, a reference electrode, and a temperature-compensating mechanism. The electrical potential created by hydrogen ion activity is measured directly by the pH-sensitive device, the reference electrode measures the ambient electrical potential, and the temperature cell compensates for any influences of temperature. Although pH measurement is a frequent bench-top laboratory function, operators also use pH probes directly in a process to monitor and control efficiency, especially in softening plants.

Chlorine Residual

Chlorine residual analyzers and recorders, both of which are very important to the operator, play a dual role in water treatment. As controlling devices, they use a closed-loop feedback system to control chlorine dose in close tolerances. As recorders, the use of which is regulated, they provide a record of the chlorine residual leaving the plant and are designed to notify the operator of pending problems with low or high amounts of chlorine.

Two electrochemical principles are used in chlorine residual controllers: amperometric and polarographic. An amperometric device uses a reagent to maintain a constant pH and an electrode to measure electrical current that is proportional to the chlorine residual. Free residual or total residual is measured using buffer solutions. Electrodes are susceptible to interferences from manganese, iron (Fe), nitrite, and chromate. A polarographic sensor contains a chlorine gas–permeable membrane. As chlorine traverses the membrane, it causes an electrical current that is proportional to the chlorine residual to be generated. This current is measured and displayed. Operators need to have ready access to this equipment while making rounds.

Analyzers that use colorimetric methods (N,N-diethyl-1,4-phenylenediamine sulfate) are also available. The systems use standardized color indicators and the intensity of the color to determine the chlorine concentration.

Chlorine residual analyzers and recorders are often used to control chlorine dosage based on the residual measure.

Total Organic Carbon

Total organic carbon (TOC) is measured by first isolating the inorganic carbon, which is then converted to carbon dioxide (CO_2) and combusted. The carbon that remains after combustion is organic.

There are several major types of instrument technologies used for on-line TOC analyzers, including the following:

- High temperature catalytic/combustion oxidation
- Supercritical water oxidation
- Patented two-stage advanced oxidation process using hydroxyl radicals
- Heated persulfate ultraviolet oxidation

Other On-Line Chemical Analyzers

On-line analyzers have been developed for many other chemicals in water. The accuracy and reliability of each must be evaluated for the intended purpose. Also, some of these analyzer test methods are not approved for compliance testing. Before purchasing an on-line analyzer, the operator should pilot-test the unit being considered under real plant conditions.

Iron, manganese, and phosphates are among the chemicals that are most often tested using on-line continuous monitors. The test methods are usually either colorimetric or a type of selective electrode. The units must be calibrated and operated according to manufacturer's instructions to give the best results.

BIBLIOGRAPHY

AWWA. 2001 (3rd ed.). M2, *Instrumentation and Control.* Denver: AWWA.

Burlingame, G.A., M.J. Pickel, and J.T. Roman. 1998. "Practical Applications of Turbidity Monitoring." *Journal AWWA,* 90(8):57-69.

Crittenden, J.C., R.R. Trussell, D.W. Hand, K.J. Howe, and G. Tchobanoglous. 2012 (3rd ed.). *MWH's Water Treatment: Principles and Design.* New York: Wiley.

Government of Newfoundland and Labrador, Department of Environment and Conservation, Water Resources Management Division. 2005. "Section 6, Instrumentation and Control." In *Guidelines for Design, Construction, and Operation of Water and Sewerage Systems.* St. John's, Nfld.: Government of Newfoundland and Labrador, Department of Environment and Conservation.

Letterman, R.D., S. Viswanathan, and J. Dwarakanathan. 2001. *A Study of Low-Level Turbidity Measurements.* Denver: AWWA and WRF (Water Research Foundation).

McTigue, N.E., M. LeChevallier, H. Arora, and J. Clancy. 1998. *National Assessment of Particle Removal by Filtration.* Denver: AWWA and WRF.

Routt, J.C. 2001. Personal communication.

Water Supply Committee of the Great Lakes–Upper Mississippi River Board of State and Provincial Public Health and Environmental Managers. *Recommended Standards for Water Works.* 2012. Albany, N.Y.: Health Research Inc., Health Education Services.

Chapter 14

Safety and Security Practices

TREATMENT PLANT SAFETY

The discussion of plant safety presented in this chapter is not intended to be a comprehensive guide on the subject. It is the responsibility of utility personnel to follow safety regulations and to be properly trained to perform their duties safely. The bibliography at the end of this chapter provides additional information on safe working practices. The information provided in this chapter should be adequate for most operator certification test examinations.

Safety and safe working conditions are the responsibility of each individual associated with a water utility. The degree of responsibility varies among upper management, supervisors, and operations personnel, but all play a role, and all are responsible for their individual actions. Safety regulations guide those actions, but the creation of policies, procedures, and methods to meet safety requirements should be a collaborative effort within the water utility.

Utility managers are responsible for maintaining safe working conditions and the physical condition of their facilities and supporting policies that encourage the safe performance of work duties. Supervisors at all levels directly control all work conditions and are responsible for the activities of the personnel they supervise. Each supervisor, foreman, crew chief, or lead staffer is responsible for ensuring that all work is performed in compliance with safety practices, utility policies, and regulations.

Employees have a particular responsibility when it comes to safety. That responsibility is to correctly use the safety equipment provided and follow all safety policies and procedures. All employees must be aware of the safety requirements and help guard against unsafe acts and conditions.

SAFETY REGULATIONS

The primary reason for developing and maintaining safe working conditions and practices is to eliminate injuries. Beyond the personal cost to an injured employee are the costs to the utility from lost time, medical expenses, and possible legal judgments. Other considerations include damage to equipment and property, the resulting repair costs, as well as hiring and training new employees to perform the work duties of the injured employee.

Occupational Safety and Health Administration Act

Another reason for developing safety policies and procedures is the federal Occupational Safety and Health Administration (OSHA) Act (Public Law 91-596). Passed in 1971, the act established OSHA and compiled numerous safety and health standards that are applicable to every industry. Specific standards have been developed for most work activities and chemical substances that an employee may be exposed to during his or her employment. These are minimum standards that must be followed; their requirements are itemized in the *Code of Federal Regulations* (CFR) (29 CFR, 1910 and 1926) (OSHA 1979, 1974). The act provides for monetary penalties, which are escalated depending on the seriousness of the safety violation, and the potential for incarceration if OSHA safety standards are violated.

If a specific requirement does not exist, OSHA relies on the general duty clause, which requires each employer to provide employment that is free of recognized hazards that are likely to cause death or serious physical harm. The general duty clause also requires employees to comply with safety practices and procedures. In addition to OSHA, many states and cities have their own safety requirements. In most cases, these standards mirror the OSHA standards; however, in some cases, state or local requirements are more stringent and carry additional penalties for failing to comply.

CONFINED SPACE RULES

A common hazard encountered in the water industry is confined spaces. Examples of confined spaces in the water and wastewater industry include access holes for valves, meters, and air vents; chemical storage tanks or hoppers; wet wells; digesters; sedimentation basins; filters; and reservoirs.

One of the most sobering statistics related to confined space safety is one relating to the death rate for those entering confined spaces. Although the death rate relating to confined spaces continues to decrease, the percentage of those who die while attempting a confined space rescue remains roughly the same. Each year, about two-thirds of those who die in confined spaces are would-be rescuers. The cause of death is usually asphyxiation or the result of an atmospheric hazard that could not be seen.

Under OSHA regulations, there are two basic types of confined spaces—permit required and nonpermit required. For a location to be classified as a confined space, the following three criteria must be met:

- The space must have limited means of entry and exit.
- The space must not be designed for continuous human occupancy.
- The space must be of a size and configuration that humans can enter to perform work.

If all three criteria are met, it is necessary to assess the space and determine if the space is a permit-required confined space. A permit-required confined space meets one or more of the following criteria:

- The space contains, or has the potential to contain, a hazardous atmosphere.
- The material within the space has the potential to engulf the entrant.
- The internal configuration is such that it could trap the entrant (i.e., downward sloping and converging floors).
- The space contains any other recognized serious health or safety hazard.

If the only hazard is atmospheric and the hazard can be controlled with ventilation, the space may be reclassified as nonpermit required. Figure 14-1 is a decision tree that can be used to identify the need for permitting a confined space.

A written confined space program is required when it is necessary to enter permit-required confined spaces. This program must:

- contain a mechanism for identifying and controlling the hazard,
- have a written entry permit,
- provide for the identification and labeling of confined spaces, and
- provide employee training, among other requirements.

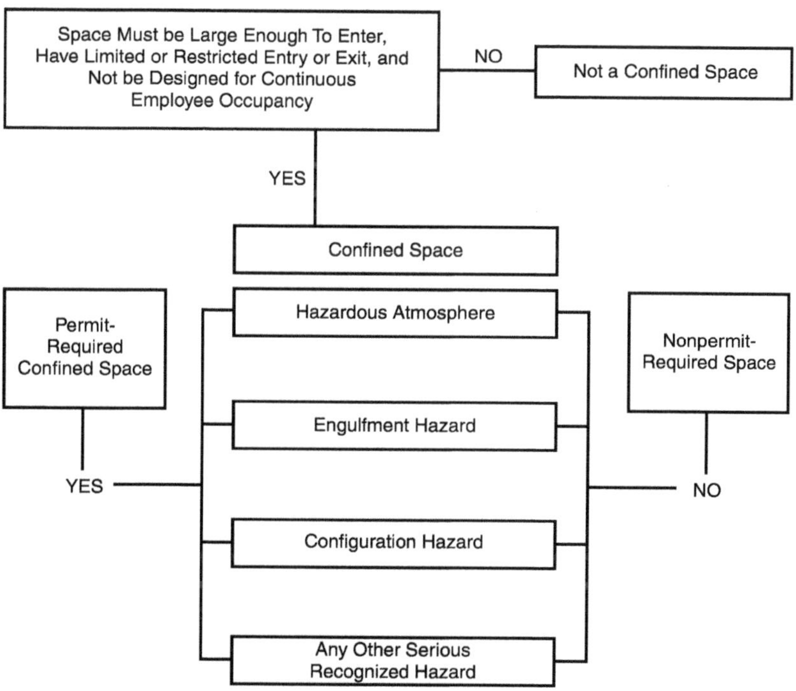

Figure 14-1 Confined space decision tree

A key component of this program is the requirement that before an employee enters a permit-required confined space, the internal atmosphere of the space be tested with a calibrated, direct-reading instrument capable of measuring oxygen (O) content, the presence of flammable gases and vapors, and potentially toxic air contaminants. In addition to these preentry tests, the atmosphere must be retested while work is being performed inside the space to ensure that acceptable conditions are being maintained. More details regarding the required program can be found in 29 CFR 1910.146 (OSHA 1979).

An employer is required to provide its employees with the specialized equipment that is required for confined space entry. This equipment includes the following:

- Testing and monitoring equipment
- Ventilation equipment (to eliminate or control atmospheric hazards)
- Communications equipment, as necessary
- Personal protective equipment (PPE)
- Lighting equipment, as necessary
- Barriers and shields to prevent unauthorized entry, as necessary
- Ladders for safe entry and exit
- Fall-prevention equipment, which is required because of the difference in elevation
- Rescue and emergency equipment, including harnesses and hoists, self-contained breathing apparatus (SCBA), stretchers or backboards, and supplies

There are three classes of employees as they relate to confined spaces: the entry supervisor, the authorized entrant, and the authorized attendant.

The *entry supervisor* is responsible for knowing the conditions within the confined space, verifying that all equipment and procedures are in place before entry, verifying the availability of rescue services and the means of summoning them, terminating entries, cancelling permits, and determining that acceptable conditions as specified in the permit continue for the duration of the entry.

The *authorized entrant* is responsible for knowing what hazards are to be faced, recognizing the signs and symptoms of exposure and understanding the potential consequences, knowing how to use any needed equipment, communicating with the attendant as necessary, alerting the attendant when a warning symptom or other hazardous condition exists, and quickly exiting when ordered or alerted.

Authorized attendants are responsible for being aware of the confined space hazards and the behavioral effects of exposure; maintaining a count and the identity of the authorized entrants; preventing unauthorized entry; remaining outside the space until relieved; communicating with the entrants and monitoring activities inside and outside of the permit space; and ordering exit, summoning rescuers, and performing nonentry rescue as required.

GENERAL PLANT SAFETY

Right-to-Know and Material Safety Data Sheet

To follow safety practices and procedures that will protect an employee and fellow workers, an employee must be aware of the hazards to which he or she is exposed. A primary method of doing this is through a Right-to-Know program developed by the employer. The main premise of a Right-to-Know program is that an employee has the right to be advised of the hazards associated with the chemicals and materials that they work with or that exist nearby in the course of their daily job activities.

Chemicals, as defined under OSHA's Occupational Safety and Health Standards (29 CFR 1910.1200) (OSHA 1979), include those used for water treatment, lab procedures, cleaning, and maintenance applications. A material safety data sheet (MSDS) (see Appendix A) lists a chemical's identity, composition, and type or types; the hazard(s) associated with its use (whether it is flammable, an oxidizer, poisonous); how it enters the body; the effects of exposure; health effect (short term, long term); the permissible exposure limit; how to handle spills or releases; and other related information. An MSDS for every chemical that is used in a water treatment plant (WTP) should be kept in an organized fashion in a location that is readily accessible to all employees.

Employees are to be trained on the hazards of the chemicals they use or are exposed to and in the methods used to protect themselves from those hazards. Refresher training should be completed by all employees at least annually.

Risk Management and Emergency Response

The US Environmental Protection Agency (USEPA) issued the Risk Management Rule on July 20, 1996, to protect public health and safety. OSHA (1979) had previously released the Process Safety Rule in February 1992 (29 CFR 1910.119), which focused on employee health and safety but did not address the safety of nonemployees. According to the Risk Management Rule, employers who store or use chemicals in quantities greater than the threshold quantity, which is defined by USEPA (e.g., more than 1,500 lb of chlorine [Cl]), must develop a written Risk Management Plan and release scenarios, conduct hazard assessments, determine means and methods to limit public exposure should a spill or release occur, and provide the public with information about their plan.

Emergency response at a water treatment facility in most incidents is responding to a chemical release. This response is governed under the Risk Management Rule and under OSHA (29 CFR 1910.1) (OSHA 1979). Several decisions must be made regarding emergency response, the first of which is to determine who will respond in the event of a chemical release. If the utility elects to have their employees respond to chemical releases, the utility is required to comply with the requirements of OSHA's Hazardous Waste Operations and Emergency Response (HAZWOPER) regulations.

Compliance requires the employer to provide specialized chemical response suits and respirators, medical exams to determine the fitness of the employees designated as responders, spill-containment materials, and annual training. Initial training is usually 24–40 hours, depending on the responsibilities the response team is expected to assume; an annual 8-hour refresher course is also required. If a utility chooses to have others act as the designated emergency response team (i.e., local fire department), the HAZWOPER regulations require awareness-level training for those who may be exposed during a chemical release. This awareness-level training is an 8-hour course with an annual refresher that focuses on the hazards of the chemicals used and evacuation procedures, routes, and staging areas to be used.

A key component of any emergency response plan is for the responder to be aware of the hazards of the chemicals that may be released and what precautions must be taken to ensure the safety and well-being of all designated responders. This information is usually provided on the specific MSDS or as part of the facility's risk management or hazard communications program.

Standard practice for emergency response is to follow the incident response system, first developed by fire and safety forces, and designate someone in charge of the emergency site. Red (hot), yellow (decontamination), and green (clean) zones should be designated, and strict controls are used to ensure that no unauthorized or unprotected entrant enters the chemical-release area. A spill or release area should never be entered alone; one standby person, fully suited in the appropriate level of PPE, should be assigned to each entrant into the hot zone.

First Aid

WTPs often provide first-aid stations for their employees. These stations are stocked with emergency supplies that are used when operators sustain minor injuries on the job. The stations, as shown in Figure 14-2, may include oxygen, bandages and antiseptics, fire blankets, and eyewash solutions. SCBA equipment may also be stored here if the location is near the chlorine area. Be sure to document use of all supplies immediately after use.

Chlorine Safety

Chlorine safety is a subset of chemical safety at many WTPs. Chlorine is a toxic chemical and is an irritant to the eyes, skin, mucous membranes, and respiratory system. Effects of exposure generally are evident first in the respiratory system and then in the eyes. The effect of exposure to chlorine is dependent on both concentration and time. The very young, the elderly, and people with respiratory problems are most susceptible to chlorine's effects. As the duration of exposure or the concentration increases, the affected

Figure 14-2 First aid and safety station

individual may become apprehensive and restless, with coughing accompanied by throat irritation, sneezing, and excess salivation. At higher levels, vomiting associated with labored breathing can occur. In extreme cases, difficulty in breathing can progress to the point of death through suffocation. An exposed person with a preexisting medical or cardiovascular condition can have a more severe response.

Anyone exhibiting these symptoms should see a qualified health care provider immediately as his or her condition will deteriorate over the next few hours. The physiological effects from exposure to various levels of gaseous chlorine are presented in Table 14-1.

To protect employees who work with or around chlorine, the employer must ensure that chlorine equipment is kept in safe working order, regular maintenance is performed on all chlorine equipment and the equipment used to handle the containers/cylinders such as overhead cranes, and employees have the proper tools and PPE.

Chlorine cylinders of 150 lb are equipped with fusible plugs that will melt at temperatures between 157 and 162°F and release the contents of the cylinder. Consequently, cylinders must be stored in a temperature-controlled area. Labels identifying the contents of any cylinder or tank car should be legible and prominently displayed. Placards with Department of Transportation (DOT) codes identifying the contents and their hazards are to be posted on the exterior of chlorine rooms so that emergency response teams can identify and address the hazards they will be exposed to in the event of a release. All cylinders should be secured and protected from accidental contact by equipment or vehicles.

Table 14-1 Physiological effects of chlorine exposure

Exposure Level *ppm*	Effects
0.02–0.2	Odor threshold (varies by individual).
<0.5	No known acute or chronic effect.
0.5	ACGIH (American Conference of Governmental Industrial Hygienists Inc.) 8-h time-weighted average.
1.0	Occupational Safety and Health Administration ceiling level.
1–10	Irritation of the eyes and mucous membranes of the upper respiratory tract. Severity of symptoms depends on concentrations and length of exposure.
3	ERPG-2 (Emergency Response Planning Guidelines as values developed by the American Industrial Hygiene Association) is the maximum airborne concentration below which it is believed that nearly all individuals could be exposed for up to 1 h without experiencing or developing irreversible or other serious health effects that could impair an individual's ability to take protective action.
10	Immediately dangerous to life and health (per the National Institute of Occupational Safety and Health).
20	ERPG-3 is the maximum airborne concentration below which it is believed that nearly all individuals could be exposed for up to 1 h without experiencing or developing life-threatening health effects.

Source: The Chlorine Institute 2023

PPE

OSHA requirements governing PPE are found in 29 CFR 1910 subpart I (see Appendix E in the CFR) (OSHA 1979), which also references various American National Standards Institute (ANSI) guidelines. As a general rule, PPE is to be considered a last resort. The elimination of the hazard through engineering changes is the optimal solution, followed by administrative changes. If these methods do not eliminate or minimize the hazard to the appropriate level, PPE is to be used. The PPE must be appropriate for the hazard it is being provided to protect the employee against. This can only be determined through a hazard assessment and equipment evaluation. In general, PPE is provided by the employer, unless it is used voluntarily or by an employee outside the work environment. In addition to providing the equipment, the employer is also responsible for ensuring that all PPE is maintained in a sanitary and reliable condition.

PPE can be provided for nearly every portion of the body. Occupational foot protection or safety shoes should be used when there is the potential for injury to the foot from falling, rolling, or piercing objects or when extreme thermal conditions exist. Different types of shoes or devices provide different types of foot protection. There are steel toe caps, both internal and external to the shoe; metatarsal guards; steel shanks; and various combinations of these devices. They are commonly worn in machine shops and during construction activities.

Another common piece of PPE is the hard hat, which is required when there is the potential for injury to the head from impact or falling objects. In a WTP, hard hats are usually worn when an overhead crane is in use or during construction projects. They may be required if there are low vertical clearances and an employee must crouch or duck to travel through the area, such as in a pipe gallery or subbasement location. There are also specific types of hard hats designed to provide protection from high-voltage shocks and burns.

Treatment plants typically have high-noise areas, usually on pump floors or in machine shops. If the employer is unable to lower the noise levels or limit employee exposure, hearing protection must be provided. Hearing protection comes in various forms, from earplugs to muffs, and offers varying degrees of protection. Noise-level sampling is required to determine what level of noise reduction is needed, which will be marked on the protective devices as a noise reduction reading. As a general rule of thumb, if an employee must raise his or her voice to be heard in the area in question, hearing protection is required. Exposures to noise levels greater than an 8-hour time-weighted average of 85 dB_A (decibels measured on the A scale) require a hearing-conservation program, which includes monitoring. Specific requirements are found in 29 CFR 1910, subpart G (OSHA 1979).

Eye and face protection may also be required when handling chemicals or doing work that produces flying debris. Impact-resistant eye and face protection (glasses, goggles, face shields) is required if flying debris is likely to be present. Chemical splash–resistant and/or vapor-resistant equipment is required for chemical handling, as specified on the chemical MSDS. Body protection, such as aprons or coats, may also be required for handling chemicals. Again, refer to the MSDS for the appropriate type and composition of protective device to be selected. The employer is responsible for providing the equipment and ensuring that it is maintained. The employee is responsible for using eye protection as required.

SCBA

SCBA is a type of respirator that supplies safe, grade D (or better) breathing air to the wearer. They are commonly used in the water and wastewater industries for emergency response or for employee protection when maintenance activities can reasonably be expected to result in the release of a hazardous material. They are also required in oxygen-deficient atmospheres or where the existing atmosphere is immediately dangerous to life and health. The other type of respirator that supplies breathing air is an airline respirator. They are not commonly used for emergency response because the length of the airline supplying the breathing air is only 300 ft. This type of respirator will not be discussed in this section. The major components of an SCBA include an air tank, harness for wearing the tank on the back, pressure regulator, hoses, face mask, and a low-air alarm; all components require routine inspection and maintenance. Refer to the manufacturers' literature

Photo courtesy of Conneaut OH Water Dept

Figure 14-3 Safety chain securing chlorine cylinders in storage

for specific maintenance intervals. DOT also has testing, labeling, and maintenance requirements for air cylinders. OSHA inspection requirements mandate inspection during routine cleaning if respirators are worn routinely and at least monthly and after each use when used for emergency purposes.

OSHA requires employees to be medically fit tested to wear an SCBA. Wearing an SCBA puts additional strain on the human body; therefore, it is important to ensure that there are no underlying medical conditions that would limit an employee's ability to use the device. An employee must be medically evaluated before using an SCBA and on a periodic basis thereafter. Annual medical exams are no longer required, but annual review of an employee's fitness by a medical provider should be performed.

In addition to medical surveillance, OSHA requires fit testing of tight-fitting respirators. The two types of fit testing are qualitative and quantitative. Qualitative fit testing involves determining whether the employee can detect the odor of test media. Irritant smoke and banana oil are commonly used for this type of test. Qualitative fit testing is subject to the impression of the wearer and may not always yield an accurate result. Quantitative fit testing involves using a probed face mask identical in brand and size to the one assigned to the employee. The testing device measures the particles in the air outside the mask and the particles inside the mask and compares the levels to determine an acceptable fit. Several manufacturers produce these testing devices. If an appropriate fit cannot be obtained with the assigned mask, another brand or size of respirator face mask may be needed to provide the required level of protection. Specific requirements regarding fit testing can be found in 29 CFR 1910.1050 (see Appendix E in the CFR) (OSHA 1979).

Hazardous Energy Control (Lockout/Tagout)

Before any employee performs any servicing or maintenance on a machine or equipment where the unexpected energizing, startup, or release of stored energy could occur and cause injury, the machine or equipment should be isolated from the energy source and rendered inoperative.

LAB SAFETY

General safety rules for laboratory workers or those who use chemical reagents in their duties can be found in the latest edition of *Standard Methods for the Examination of Water and Wastewater* (2022). All WTP employees should be familiar with these general rules. The previous discussion on PPE applies particularly to laboratory workers.

GENERAL SAFETY

WTP structures can present the need to implement safety practices, and operators are urged to use a commonsense approach to identify these needs. Operators should work with designers of their treatment plant structures and workspaces (see Figure 14-4) to ensure safe equipment and accommodations are made a part of the overall safety plan.

PLANT SECURITY

Water Utility Security Initiatives

The Department of Homeland Security has taken steps in partnership with all public and private stakeholders to ensure the protection and resilience of water systems. These water sector partners collaborate to be better prepared to prevent, detect, respond to, and recover from terrorist attacks, other intentional acts, and natural disasters, otherwise known as the *all-hazards approach*.

Homeland Security Presidential Directive 7 (CISA 2003) identifies 18 critical infrastructure and key resources sectors and assigns an agency lead for each. The USEPA is the designated agency for the water sector.

Photo courtesy of Aqua Ohio Struthers

Figure 14-4 Fire protection Equipment at Aqua Ohio Struthers WTP

All-Hazards Approach

Water utilities have taken steps to protect critical water supply facilities from theft and sabotage and have written plans for response and recovery from events that interrupt water service. An all-hazards approach to security and emergency preparedness mirrors the time-tested multibarrier approach used in water treatment.

The Water Sector Coordinating Council consists of representatives from a cross section of water agencies and professionals. The council interacts with federal agencies regarding homeland security issues. In this role, the council has developed several critical resources for water utilities:

- *Water Sector–Specific Plan: Annex to National Infrastructure Protection Plan* (USEPA and DHS 2015)
- *All-Hazard Consequence Management Planning for the Water Sector* (Preparedness, Emergency Response, and Recovery CIPAC Working Group 2009)
- *Roadmap to Secure Control Systems in the Water Sector* (Water Sector Coordinating Council Cyber Security Working Group 2008)
- *Water Sector Decontamination Priorities: Recommendation and Proposed Strategic Plan* (Water Sector Decontamination Working Group, CIPAC 2008)

In addition, standards have been developed as part of the AWWA/ANSI process to assist water systems in their efforts to improve facility security and emergency preparedness:

- ANSI/AWWA Standard G430, Security Practices for Operations and Management
- ANSI/AWWA Standard G440, Emergency Preparedness Practices
- ANSI/AWWA Standard J100, Risk Analysis and Management for Critical Asset Protection (RAMCAP) Standard for Risk and Resilience Management of Water and Wastewater Systems

WTP operators should be aware of the requirements for the storage and transportation of hazardous substances. Specifically, chlorine gas and large quantities of sodium hydroxide (NaOH; caustic soda) have extensive requirements for emergency response and security, called risk management plans. Many treatment plants have elected to abandon using chlorine gas because of these requirements and concerns. Onsite generation of sodium hypochlorite (NaOCl) has gained popularity to avoid these hazards.

Treatment Plant Security Measures

Protecting a public utility's assets to ensure an uninterrupted, safe water supply is a fundamental concern of all levels of management. Plant operating staff should participate in the development of a master plan that will establish a level of protection for their areas of responsibility. A utility should implement a comprehensive master plan to provide for overall security for the utility. These plans are site specific but contain common components. By breaking down these components into areas of responsibility, a utility can assure an integrated approach to plant security development. The major features of a security master plan are described in the following paragraphs.

Figure 14-5 Decorative security fence

Physical Requirements

The physical requirements of the plan focus on actions that terrorists, saboteurs, or criminals might take to damage or contaminate the water supply or sabotage system infrastructure and the electric transmission system. Preventing acts of sabotage at WTPs, electric substations, maintenance facilities, other utility facilities, and in the distribution system requires physical barriers to prevent intruders from entering a site and reaching vital areas or equipment.

Integrated Systems Approach

An integrated systems approach that combines people, equipment, and procedures is used to protect against threats and intrusions. The security plans have multiple layers of protection that provide several methods to monitor sites and provide deterrence measures for a multitude of threats. It also monitors employee activities and documents the presence of contractors, visitors, and others at the facilities with swipe cards and proximity card readers.

Detection Systems

Electronic detection systems provide alarms to alert security and plant personnel that the plant or facility is being approached or an intrusion attempt is under way. Many cost-effective devices are available to detect movements, forced entry, and even record any events and times. Video surveillance is another useful detection system that can aid operators to prevent unlawful entry to the plant facility.

Delaying Tactics

Barriers, such as chain-link fences around perimeters, gates, and proximity card readers, are arranged in layers to create physical and psychological deterrents and to delay intrusion long enough for security to detect the intrusion and respond. Barriers channel entry and exit through specified control points, facilitating efficient identification and control of authorized people and equipment.

Assessment Systems

Closed-circuit television cameras monitor facility entrances and special secured areas and allow operators to assess an alarm that was received while limiting exposure until assistance arrives. These systems can be linked with other security systems to control entry and record images for use by authorities.

Communications and Response Systems

Security assessment and monitoring devices are linked to communications systems to transmit information to a local monitoring site, a central monitoring site, and/or the security force. As the security officer or facility personnel respond, they should remain in communication with a central monitoring site and other site personnel at all times by several means of alternate communication, such as 800-MHz radios, cell phones, walkie-talkies, pagers, and other communication devices available at the site.

Training and Qualifications

Security and facility personnel effectiveness must be maintained through training. Training exercises against mock adversaries should be conducted periodically to add realism to the training. In addition, training on system maintenance, reporting, and policy enforcement must be conducted annually for effective system management.

Access Control

Employees are issued identification badges that must be carried at all times. Access control systems at the entrance gate and within buildings will confirm a person's identity and authorization before entrance is granted. Visitors should also be issued badges after entry into the facility.

Reliability of Personnel

Security awareness will become part of each employee's initiation to the work site. Thorough background investigations should be conducted in accordance with department policies as the primary method for learning about any past actions that may indicate problems with individual employees. Background checks should be performed on all potential employees before hiring.

Testing, Maintenance, and Auditing

Periodic testing of all security equipment is required to make sure it is operating correctly. If problems are found, corrective measures are required to ensure the security function provided by the equipment is immediately restored. Physical security programs should be audited periodically to ensure certain requirements are being met.

Contingency Planning

Plant operators and facility managers must plan and provide for all reasonable contingencies, such as employee strikes, natural disasters, and onsite emergencies that require quick support and coordination from outside agencies and security personnel. Backup personnel, vendors, emergency response agencies, and others should be identified to respond to these unusual situations. These plans should be reviewed annually to ensure that contact information is current.

BIBLIOGRAPHY

ANSI/AWWA. 2009. G430. Security Practices for Operation and Management. Denver: AWWA.

ANSI/AWWA. 2011. G440. Emergency Preparedness Practices. Denver: AWWA.

ANSI/AWWA. 2010 (R13). J100. Risk Analysis and Management for Critical Asset Protection (RAMCAP) Standard for Risk and Resilience Management of Water and Wastewater Systems. Denver: AWWA.

AWWA. 2018 (5th ed.). M19, *Emergency Planning for Water Utilities*. Denver: AWWA.

AWWA. 2002 (6th ed.). M3, *Safety Practices for Water Utilities*. Denver: AWWA.

AWWA. *Utility Security*, website resources. Denver: AWWA.

CISA (Cybersecurity & Infrastructure Security Agency). 2003. "Homeland Security Presidential Directive 7." www.cisa.gov/news-events/directives/homeland-security-presidential-directive-7 (accessed 2025)

The Chlorine Institute. 2023 (1st ed.). *Water and Wastewater Operators' Chlorine Handbook, Pamphlet 155*. Washington, D.C.: The Chlorine Institute.

Meckes, T.D. 2024. Personal communication.

OSHA (Occupational Safety and Health Administration), Department of Labor. 1979. "Safety and Health Regulations for Construction." 29 CFR 1926. *Federal Register*, 44:29:8577.

OSHA, Department of Labor. 1974. "Occupational Safety and Health Standards." 29 CFR 1910. *Federal Register*, 39:125:23502.

Preparedness, Emergency Response, and Recovery CIPAC (Critical Infrastructure Partnership Advisory Council) Working Group. 2009. *All-Hazard Consequence Management Planning for the Water Sector*. Washington, D.C.: CIPAC.

Raimann, M. 2001. Personal communication.
Standard Methods for the Examination of Water and Wastewater. 2022 (22nd ed.). Washington, D.C.: APHA, AWWA, and WEF.
Sundheimer, M. 2001. Personal communication.

Chapter 15

Record Keeping, Reporting, and Data Management

Records of plant operations and maintenance activities provide a basis for predictive decision-making and proof of compliance. Records also help operators to efficiently use chemicals (minimize waste) and equipment.

Reporting is typically the responsibility of the plant manager, but the lab staff also may be required to report. Each plant should have a standardized format for reporting internally, and of course will need to know how to use the regulator's reporting system. The monthly operating report (MOR) that is sent to the regulator's office is mostly done online, so the water treatment plant (WTP) staff needs an Internet connection and a good understanding of cybersecurity protocol.

Data are something all WTPs generate and are a source of information that is often neglected. Data can provide important information and clues about the operational habits of the staff that can and should be exploited by the manager. WTP optimization efforts usually depend heavily on the ability to sift through the data and make sense of them.

As discussed elsewhere in this book, data management has become a huge problem for WTP operators who are unable to retrieve the data they need for analysis and for reporting in special cases because of vendor-supplied supervisory control and data acquisition (SCADA) systems.

PROCESS RECORDS

In addition to the records that are required by the local regulatory agency, a WTP should maintain records of activities as they relate to each process or discipline. Refer to the AWWA (2010) *Principles and Practices of Water Supply Operations* series for additional information on this topic.

The following sections describe the types of records operators should keep when performing their duties.

Coagulation and Flocculation

Maintain records of past performances for coagulants, including dosages, turbidities (sedimentation basin and filtration), particle counts, filterability indexes, zeta potentials, total organic carbon and ultraviolet absorbance at 254 nm, as well as the temperatures, pH values, and color values at which the coagulants performed. For softening plants, record which softening chemicals are used (see Chapter 9). Also record detention times and overflow rates in basins used for softening as well as residual quantities produced. Keep equipment records such as hours run for each mixer and flocculator, basin-cleaning activities, maintenance performed on motors and gears, and electrical use records.

Sedimentation

Calculate and record surface overflow rate, weir overflow rate, turbidity performance of each basin, and quantity of sludge removed from each basin (with a calculation for percent solids).

Maintain records of any operations and maintenance activities for each unit, including blowdown, cleaning, raking, and so on. Safety records for confined space entry may be necessary.

The Partnership for Safe Water suggests that the turbidity performance (and particle count, if available) in each sedimentation basin be recorded periodically (daily, every 4 hours, or hourly). Record the settled water turbidity as a combined flow but realize that it cannot characterize the performance of an individual basin.

Filtration

For each filter, record on a form as shown in Figure 15-1 the rate of flow, in million gallons per day or gallons per minute; head loss; length of run and unit filter run volume (UFRV); turbidity (in accordance with the Interim Enhanced Surface Water Treatment Rule [IESWTR]); particle counts; and dosage of any filter aid used. At each backwash, record the amount of backwash water and surface wash water used, the length of the backwash, any observations during backwash, and time of filter-to-waste if used. Record the percentage of water treated that the backwash water represents (make note of amounts greater than 4%). All information can be included on one form.

On a quarterly basis, an operator should record bed expansion measurements, condition and depth of media, and an evaluation of the bed surface. On a yearly basis, record the results of solids retention analysis before and after backwash and the results of sieve analysis of the media. Record the addition of any media to the bed, including amounts and types. Keep a record of underdrain inspections and any work that was done on any appurtenances.

Figure 15-1 A sample filter log sheet

Time	Rate of Flow *mgd or gpm*	Head Loss *ft*	Hours in Service	Total Volume Filtered (Cumulative)	UFRV *gal/ft²*	Turbidity *ntu and particle counts*	Backwash Start	Length of Backwash and Amount Used *min and gal*	Surface Wash *amounts and min*	Filter Aid Applied, Dosage	Filter-to-Waste at Startup *ntu*
0000											
0200											
0400											
0600											
0800											
1,000											
1,200											
1,400											
1,600											
1,800											
2,000											
2,200											

UFRV—unit filter run volume

Source: Nick Pizzi

Record results of any testing required for filter exceptions reporting under the IESWTR (this is regulated). Many states have required forms for plant operators to submit for filter performance. Consult the state regulatory agency for specific requirements.

Chlorination or Disinfection

Maintain records of the type of disinfectant used, including ordering information (phone number, address, and shipment amounts and container types), as well as information on costs and current dosage rates. Also, results of bacteriological testing should be recorded for each process that uses a disinfectant, the water temperature and pH, and any unusual conditions that may occur. Keep records of safety training.

Precipitative Softening

Record the amount of softening and recarbonation chemicals used and the results of jar testing and lab analyses that justify these amounts. Record the results of analysis for alkalinity, pH, hardness, and related parameters for each shift, preferably every 2 hours for the source water and for all steps in the process. Keep a record of the chemical feeder setting and record the amount of water treated. Put everything on one form and maintain it daily.

Record the amounts of sludge produced. If a solids-contact unit is used, record the sludge blanket depth and results of settling tests. Keep records of the amounts and rates of solids returned to the process, if applicable.

Ion Exchange

Record the hardness, alkalinity, magnesium, calcium, and pH of the source and treated waters. Record the total amount of water treated and the amount of water treated by each softening unit, including bypass water. Also, keep a record of the amount of salt used for regeneration and the length of each cycle between regenerations. Record the amount of backwash water used and the amount of waste sent to sewer or other disposal places.

Aeration

Keep process records of source water and finished water quality to determine if the process is accomplishing its goals. Also, record the daily quantity of water treated, the details of safety and maintenance procedures, and changes in other treatment processes that may be affected by aeration. Document the process operating conditions, including the air-to-water ratio, to calculate performance and cost information.

Adsorption

Records kept regarding activated carbon differ depending on the type of carbon used:

- *Powdered activated carbon (PAC)*: Record the types of tastes and odors being experienced, the threshold odor number (TON), and the dates of taste-and-odor occurrences. Keep a record of the coagulant and PAC dosage predicted from jar tests as well as the actual dosage used.
- *Granular activated carbon (GAC)*: Record the number of filter hours; UFRV; any losses of GAC from the filter, measured monthly; dates of installation; and periodic TON from each filter. Also, record when backwashing was performed and the amount of backwash water used as well as the raw-to-finished organic content removal.

Iron and Manganese Removal

If using ion exchange, the previously referenced records can be used. In addition, record raw and finished iron (Fe) and manganese levels and results of distribution systems analyses performed. If permanganate is used, record amounts and dosages, including those predicted by jar testing.

If a sequestrant is used to control iron and manganese, record the amount and type used and the data from distribution system flushing efforts.

Fluoridation

Keep a daily record of the amount of fluoride in the finished water and the source water, if necessary (a weekly source water analysis may be sufficient if the regulator allows). Also, keep a daily record of the chemical dosages used and record the chemical feed rate each hour of each shift. Compute and record the daily fluoride chemical feed and compare it with the lab analysis as a double-check. Record safety classes and lectures that were provided to operators.

Corrosion Control

Maintain records of corrosion chemical dosages and any data available as to their performance, such as coupon testing results, corrosion meter readings, temperatures, flushing activities, and lead (Pb) and copper (Cu) compliance results, i.e., 90th percentile levels and maximum levels. If phosphates are used, keep strict records of dosages and amounts of total and dissolved phosphate in the system. Flush mains regularly, especially dead-end mains; sample for phosphates at 5-, 10-, and 15-min intervals into the flush; and record results. Record the main flushing velocities in feet per second. Track any customer complaints that may relate to use of corrosion chemicals.

Water Main Flushing

Main flushing records are best kept in a searchable database. Record main location and size; size of flushing pipe or hydrant; flow rate at flush; gallons used for flushing; velocity in feet per second; chlorine (Cl) residual; amount of phosphate used, if any; and results of biological testing, such as heterotrophic plate count quantity or quantity of coliform. Query the database by main size and show gallons used and time needed, along with any chemical and biological data, to generate monthly reports. In this way, a profile will emerge that shows the amount of water necessary to flush each size main, the time it takes, and the results obtained.

REPORTING

Reporting requirements are generally a function of the compliance reports that local and federal regulators require. In general, compliance testing results must be reported no later than 10 days after the end of the month in which they were accumulated. It is good practice for WTP staff to provide reports to upper management. Even if the administration does not require these reports, they should be generated and kept on file for future reference.

Like a diary, a written report is an excellent reminder of events that took place and the methods used to handle these events. Internal memos of safety issues, personnel and disciplinary actions, purchases, and maintenance events can actually help protect staff from allegations of nonfeasance. Reports should be written as soon as possible after an event takes place so that details can be easily recalled. Do not write reports or internal memos in anger; take time to cool down before recalling an event.

Most compliance reports must conform to a specific format. This enables the regulator to assemble large amounts of data from many utilities in an orderly fashion. Internal or outsourced reports, other than compliance reports to regulators, should include the following:

- Times and dates of the audit and any description of incidents or accomplishments
- Financial or inventory considerations
- Personnel involved
- Conclusions drawn, with supporting data and references
- List of report recipients, with acknowledgment of receipt, if appropriate
- An executive summary if the report is lengthy and involved

The report should be filed in a logical order, e.g., file reports under general categories such as research, safety, finance, and so on, so that the report can be easily retrieved in the future. As time permits, reread all reports to gauge progress in reaching goals and to learn from past mistakes.

Reporting Treatment Incidents

There are specific requirements for the time and method of reporting a regulatory violation. These are described in the drinking water regulations for the applicable regulatory agency. However, there are times when it is not clear if there has been a violation or when there is a water quality incident that may affect consumers. The WTP personnel should have a predetermined plan of action to deal with these situations.

The notification and action plan for incident reporting should include plant management, utility management, other utility officials, and regulatory agencies. The utility should have discussions with the regulatory agency representatives before any incident to establish an understanding of how they will handle this type of communication. The agency may have technical resources to assist the plant personnel with a response. Also, the agency may be helpful when dealing with mass communication media regarding the incident.

Plant Performance Reports

Periodic (monthly, quarterly, and annual) plant performance reports should be prepared, and these should be examined to detect trends and identify opportunities for improvement. The operating reports previously described are the basis for these performance reports. Performance reports summarize the operating results and usually include various charts and calculated performance measures. These summary reports do not need to be lengthy and, indeed, can be added to the operating reports if desired. The purpose of the performance reports is to examine the plant results over longer time periods to see if there are trends that need investigation. All plant operating personnel should review these reports to provide feedback and to participate in efforts to meet plant operating goals.

DATA MANAGEMENT

Data that are generated at a WTP are valuable. In addition to the regulatory requirements for keeping and reporting data as proof of compliance each month, the data can also be a source of information that can aid the staff in their optimization efforts. Most of the data are generated by the operational control of each process, such as continual filter effluent turbidities, hourly pH and alkalinity measurements from a softening unit, or daily contact time calculations in a clearwell. Some data are recorded on the shift operator logs, some in the memory banks of the SCADA system, and some in the lab quality assurance notebook. Much of this information is compiled and formatted into a MOR. And then, often it is promptly forgotten.

This vast reservoir of knowledge can also serve as a starting point to the optimization efforts of the WTP. Any optimization effort worth doing always starts with an assessment—a baseline—of where things stand right now versus what end point, or goal, is desired. To create a baseline and measure the amount or rate of goal achievement in an ongoing process like a WTP requires parsing or examining the data in detail. A spreadsheet is needed for this.

If your WTP data are being accumulated by a SCADA system, and you can't retrieve the data at will, you should make it clear to the SCADA vendor that you need to access your own data at will.

BIBLIOGRAPHY

AWWA. 2010 (4th ed.). *Principles and Practices of Water Supply Operations—Water Treatment*. Denver: AWWA.

Linder, K., and B. Martin. 2015. *Self-Assessment for Water Treatment Plant Optimization*. Denver: AWWA.

Pizzi, N. 1995. "Use It—Don't Lose It: How Data Can Improve Operations." *Opflow*, 21(9):1-8.

USEPA (US Environmental Protection Agency). "Reporting Requirements for Regulations." Washington D.C.: USEPA. www.epa.gov/dwreginfo/drinking-water-regulations

Chapter 16

Logic-Based Problem Solving

This chapter provides advanced problem-solving challenges to operators. A process line diagram of a water treatment plant (WTP) is shown, followed by a narrative of the operational scheme and data in Tables 16-1 and 16-2 that are needed to answer a series of questions requiring operators to use reading, math, and logic skills to think through the scenarios presented. For curious-minded operators, each problem should provoke them into asking the universal question: "Why do I need to know this?"

MODEL PLANT #1: PROCESS LINE DIAGRAM

Model WTP #1 (not to scale)

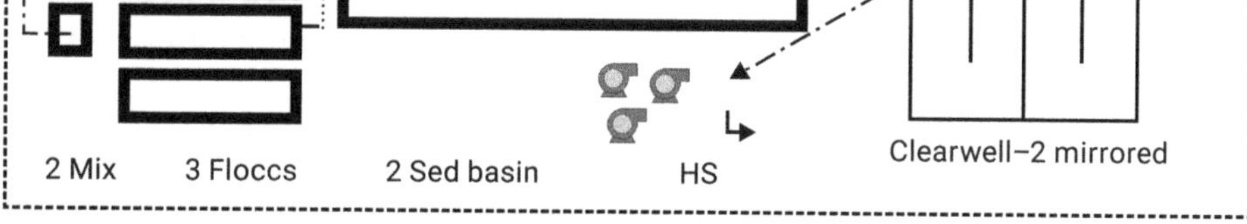

Mix—rapid mixer
Flocc—flocculator
Sed basin—sedimentation basin
HS—high service pumps
SFBW—spent filter backwash water

General Overview of the Model Plant

The model WTP was built as a 6 million gallon (MG) per day (6-mgd) conventional treatment plant that employs chemical mixing, flocculation, sedimentation, filtration, disinfection, fluoridation, pH adjustment, and corrosion control to produce drinking water. A few years ago, the WTP staff installed higher rate media configuration in the filters and tube settlers in the existing sedimentation basins. This allowed them to procure a 9-mgd rating from the regulator and keep the same footprint.

The plant obtains its source water through an intake that extends into Lake Erie. A remote pumping station is situated north of the treatment facility and houses parallel wet wells and screens that serve as suction to the source water pumps. A potassium permanganate ($KMnO_4$) feeder and day tank are located at the screen area, and the chemical is fed to the suction side of the screens for taste-and-odor control and as a preoxidant.

Source water flows into one or both rapid mixers, where liquid alum is added for total organic carbon (TOC) removal and coagulation. Water then flows to flocculation, sedimentation with tubes, and dual-media filtration processes. The water is disinfected with chlorine (Cl) ahead of the filters, and once again as it flows to the clearwell along with fluoride and phosphate. There is no way to measure total flow into the clearwells because the filter flowmeter and effluent totalizer on filter #4 are out of service, so the superintendent is using the plant high service meter and clearwell level to estimate finished water pumpage and production daily.

As shown in the process line diagram, the model plant has two rapid mixers, three flocculators, two sedimentation basins, and four dual-media filters. There is a disinfection clearwell in front of the plant that has two mirrored halves, and it is baffled (factor of 0.38). Each half is 72 ft × 72 ft (or 5,184 ft^2). The depth is 15 ft, but the operators maintain a level of about 12 ft most of the time and don't allow the depth to go below 8 ft.

A 300,000-gal wash-water storage tank sits just to the east and north of the plant. When filters are washed, the spent filter backwash water (SFBW) is captured in a 250,000-gal holding tank where it can be returned to the rapid mixer at a controlled rate when the operator is ready to do so.

There are four high service pumps in the plant that move water into the distribution system. Two are variable-frequency drive pumps, and two are fixed speed pumps.

Chemicals are fed using flow rate proportioning based on treated water flow rate through the plant, which is quantified by using the settled water flume level to determine velocity and therefore flow rate. Adding to the challenge, wash water is recycled at a rate of about 10% of the instantaneous source water flow rate as prescribed by the Partnership for Safe Water. That recycling flow, however, enters only one of the rapid mixers, not both. Additional design information is provided in Table 16-1, and operating data is in Table 16-2.

Table 16-2 provides the production values and chemicals used in the plant during September.

Table 16-1 Design information for source water system

Component	Number of Units	Dimensions/Information
Intake	One	4,000 of 48-in. main
Screens	Two	Auto traveling
Potassium permanganate feed system	One day tank, two feed pumps	Uses a 0.2-lb/gal solution
Screen wells, rectangular	Two	Bottom elevation at 558 ft, 7.5 ft × 18.5 ft each, design lake level at 573 ft
Pumps	Four	Two at 2,604 gpm at 93-ft TDH, 105 hp; one at 1,750 gpm at 93-ft TDH, 75 hp; one at 3,500 gpm at 93-ft TDH, 125 hp
Pump wells, rectangular	Two	Bottom elevation at 558 ft, 24.5 ft × 30 ft each
Main to rapid mixers at WTP	One	2,560 ft of 36-in. main

Source: Nick Pizzi
TDH—total dynamic head, WTP—water treatment plant

Table 16-2 Chemicals used and production values for the month

Chemical	Gal/lb Used	Notes
Liquid alum	5,318.67 gal	48.8% dry basis, 1.325 specific gravity
Chlorine gas	5,473 lb	1 mg/L fed B4 filters
Hydrofluosilicic acid (H_2SiF_6)	679 gal	23%/10.5 ppg
Phosphoric acid	910 gal	30%, 12.2 ppg
Potassium permanganate	467.6 lb	Made at 0.2 ppg
Powdered activated carbon	4,475 lb	Silo storage, slurry fed
Filter aid polymer	4.6 gal	Used for two days
Production		
Water treated in month	186.90 MG	Raw station totalizer
Finished water pumped per day	5.92 MG	From HS pump meter
Filter production	#1: 46.54 MG, #2: 44.70 MG, #3: 46.24 MG, #4: N/A	
Filter runs per unit	Each filter had 10 filter runs	
Total wash water used	1,785,960 gal	

Source: Nick Pizzi
MG—million gallons, Ppg—pounds per gallon, HS—high service pumps, B4—before

OPERATIONAL MATH QUESTIONS FOR THE MODEL WTP

Process question: The intake stretches out into Lake Erie, and it is designed to convey 10 mgd as a maximum sustained flow rate. Design engineers like to make sure that the velocity of a water conveyance system is swift enough to prevent premature settling, but not too swift that friction losses would be prohibitive. They generally specify that the main be sized to handle a flow of 1–5 feet per second (fps). What is the velocity in the main at the average day flow rate?

Answer: Using the formula Velocity (V) = Flow rate/Area of the pipe, the operator can determine an average velocity at the 10 mgd. The formula is usually written as $V = Q/A$. In this case, it is as follows:

$$V = \frac{10 \text{ mgd} \times 1.55 \text{ ft}^3/\text{s per mgd}}{0.785 \times 4 \text{ ft} \times 4 \text{ ft}} = 1.23 \text{ fps}$$

This calculation only provides an average flow velocity. Along the side walls where there is fluid drag, the velocity would be slower. That also means that the water flowing through the inner portion of the pipe is moving faster. Because the average day for this WTP is much lower than the max rating, the operator might expect that the low velocities will allow silt and sand to accumulate, and so frequent maintenance service from a diver is likely to be needed.

Process question: $KMnO_4$ is fed into the screen wells to begin some taste and odor treatment and to help oxidize a small portion of the TOC. It costs $3.12 per pound, and the operators make a 0.2-pound-per-gallon solution of it in the day tank. The plant superintendent is concerned about using it because she has taken half of the source water pump wells out of service for cleaning, has been using as much as 0.8 mg/L of the chemical, and wants to avoid "red water" breaking through her filters. Her pH through the pretreatment train is low and to the point where she needs at least 50 min of detention time. A higher pH would decrease the reaction time needed, but she has no caustic soda or lime. She wants the permanganate reaction to go to completion before adding coagulant to the mixers. At 6.2 mgd, does she have enough time? If so, how many mL/min needs to be fed to either increase or decrease the feed rate by 0.2 mg/L? How much is the increase going to cost each day?

Answer: The permanganate is described as fed into the area in front of the screens, so she can ignore that space and count the volume of the wet well and of the 36 in. main from the raw station to the mixers as her reaction vessel. At 6.2 mgd, the flow rate is 10.075 ft³/s (6.2 × 1.55), or 604.5 ft³/min (60 × 10.075). The volume for reaction is the volume of one wet well and the volume of 2,560 ft of the 36-in. main. Therefore, the following applies:

$$\text{Volume of wet well} = (15 \text{ ft}_{\text{(lake level – bottom level of pump well)}}) \times (24 \times 30 \text{ ft}) = 10{,}800 \text{ ft}^3$$

$$\text{Volume of water main to mixers} = 0.785 \times 3 \text{ ft} \times 3 \text{ ft} \times 2{,}560 \text{ ft} = 18{,}086 \text{ ft}^3$$

$$\text{Detention time} = (10{,}800 + 18{,}086) / (604.5 \text{ ft}^3/\text{min}) = 47.8 \text{ min}$$

There is sufficient reaction time. An increase in $KMnO_4$ feed rate is safe and may help with the pretreatment. To add another 0.2 mg/L to the existing feed rate using the day tank mixture of 0.2 lb/gal (roughly 2%) solution, the calculation is as follows:

ppd = 0.2 mg/L × 6.2 mgd × 8.34 × (1 gal/0.2 lb) =
about 10.34 lb/d increase

10.34 ppd × 453.6 g/ lb = 4,690 g/d

4,690 g/d × 1 d /1,440 min = 3.257 g/min

3.257 g is 3,257 mg, so (1 mL/20 mg) × 3,257 mg/min =
162.85-mL/min increase needed.

A 10.34 lb/day increase will cost (10.34 ppd × $3.14 per pound) =
about $32.47 per day.

Process question: If the large source water pump is 90% efficient, and the driver is 88.6% efficient, what does it cost to operate that system for 12 hours if the electricals rate is $0.145 per kW•h? Assume friction losses to be negligible.

Answer: The system is set up to pump 3,500 gpm (5 mgd) through a 36-in. main that is in good shape, so friction is not an issue, and we can simply work with the design head stated as 93 ft. The calculation is as follows:

$$\text{hp} = \frac{3{,}500 \text{ gpm} \times 93 \text{ ft}}{5{,}960 \times (0.86 \times 0.9)} = \text{about } 106 \text{ hp}$$

To convert horsepower to kilowatts (kW), we need to multiply by 0.746. The work expressed in kW is (106 × 0.746) = 79.07 each minute. We can look at it this way: 3,500 gal of water, or 3,500 gal × 8.34 lb/gal, equals 29,356.8 lbs of water are being lifted up through a height of 93 ft every minute. After 12 hours, or 720 min, the pump will have done work equal to (29,356.8 lb × 93 ft) = 2,730,182.4 ft-lb.

That works out to be about 1.02 kW•h, or about 15 cents/min. For 720 min, we would be billed for (720 × $0.15) = 10,800 cents, so the cost is *$10.80 per 12 hours of pumping.*

Process question: The water reaches the two rapid mixers that are each 2.5 ft^2 and have a depth of 7 ft. These are the original units, and they were not changed during the upgrade process from 6 to 9 mgd. Here the coagulant is added to begin the coagulation process (see Chapter 5). With one mixer in service, what is the detention time through them at 3.3 mgd?

Answer: The operators know that detention time in rapid mix is supposed to be 30 s. They have the option to operate one or both of them at any time. Originally, these mixers were rated at 3 mgd each.

Taking one out of service for cleaning or maintenance could be an issue, and so they try to plan carefully for outages. The detention time is calculated as follows:

$$3.3 \text{ mgd} \times (694 \text{ gpm}/1 \text{ mgd}) = \text{approximately } 2{,}290 \text{ gpm}$$

$$\text{dt} = \text{Vol/Flow Rate} = \frac{4.5 \text{ ft} \times 4.5\text{ft} \times 7 \text{ ft} \times 7.48 \text{ gal per ft}^3}{2{,}290 \text{ gpm}}$$

$$= 0.46 \text{ min, or } \textit{about } 28 \ s$$

Process question: In a conventional surface WTP like this one, the addition of an aluminum-based coagulant helps the staff meet the requirements of the Disinfectants and Disinfection Byproduct Rules and the Long-Term 2 Surface Water Treatment Rule. It is supposed to remove 25% of the incoming TOC and meet the turbidity provisions of the rules that state that the filtered water must be below 0.3 ntu 95% of the time and never over 1 ntu (see Chapter 1). Lake Erie source water typically needs more coagulant for the TOC removal than it does for turbidity removal. What is the dosage of alum at the 6.6 mgd flow rate if it is 48.8% dry basis, has a specific gravity of 1.325, and 482 mL per minute are being fed? How much alkalinity might be consumed?

Answer: If the specific gravity is 1.325, the solution weighs 11 lb/gal (1.325 × 8.34). At 48.8%, there are 5.36 lb (11 × 48.8%) of dry-basis alum in each gallon of alum. Therefore, it is 642 g/L (5.36 lb/gal × 453.6 g/lb × 3.785 L/gal), or 642 mg/mL dry-basis alum. If 482 mL/min are being fed to treat 6.6 mgd, the calculation becomes the following:

$$(6{,}600.000 \text{ gal/d} \times 1 \text{ d}/1{,}440 \text{ min} \times 3.785 \text{ L/gal})$$
$$= 17{,}668.75 \text{ L/min, and}$$

$$\text{mg/L Dry-basis alum dosage} = \frac{484 \text{ mL/min} \times 642 \text{ mg/mL}}{17{,}668.75 \text{ L/min}} = 17.6 \ mg/L$$

Because alum uses up alkalinity at a ratio of 0.5 mg/L alkalinity to 1 mg/L alum, *you would expect a drop of 8–9 mg/L.*

Process question: The three plant flocculators are 14 ft × 45 ft × 13.3 ft water depth each and have four compartments. There is a horizontal drive shaft with arms and paddles attached. The last compartment has two arms with two sets of three paddles each. The arms are at 180 degrees from each other. The distance from the middle of the shaft to the tip of the outermost paddle is 6 ft.

What is the average detention time at 6.2 mgd if two flocculators are in service? If the operator wants to use a final compartment paddle tip speed of 1.2 fps, how long should one revolution take?

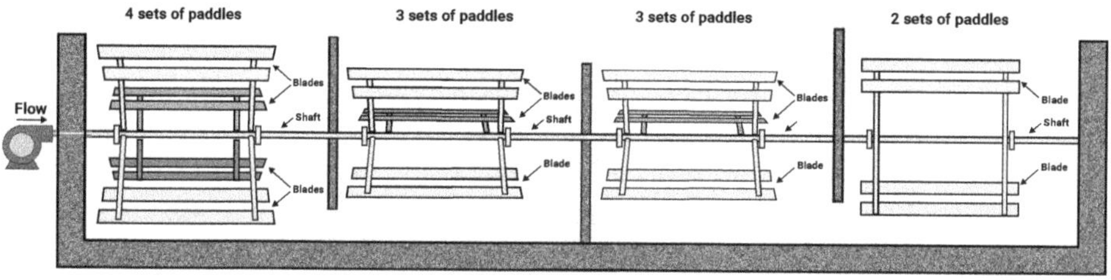

Answer: Two units in service would provide 16,758 ft³ of volume ([2] × [14 × 45 × 13.3 ft water depth] = *dt* = 16,758 ft³/[6.2 mgd × 1.55 fps/mgd]), which equals about 1,744 s, or *dt = 29 min.*

Travel distance for the outermost paddle tip would be (π × D) = 3.14 × 12 ft = 37.68 ft. At 1.2 fps, *one revolution should take 37.68/1.2 = 31.4 s.*

Process question: When this WTP was constructed, the design parameters for a settling basin were shown in the "Ten State Standards" as requiring a detention time of 4 hours or less, a surface overflow rate (SOR) of 0.5 gpm/ft² or less, a weir overflow rate (WOR) rate of 10 gpm/ft, and a flow-through velocity of 0.5 ft/min or lower (see Chapter 6). Calculate these for the basins at the design flow rate of 6.0 mgd.

Answer: These design parameters were required at the time the WTP was constructed. They were thought to be the optimal configuration of a rectangular sedimentation basin in a conventional WTP using alum as a coagulant to treat a source water like Lake Erie.

6 mgd converts to 4,164 gpm (6 × 694) and 9.3 ft³/s (6 × 1.55).

At 37.75 × 120 × 14.8 ft water depth (each), the volume would be 134,088 ft³ or about 1 MG total. Detention time with both in service therefore is as follows:

$$\text{dt} = \text{Vol / Flow Rate} = \frac{(134{,}088 \text{ cu ft})}{6 \times 1.55} = 14{,}418 \text{ s or 4 hours, or}$$

$$\text{dt} = \text{Vol / Flow Rate} = \frac{(1{,}000{,}000 \text{ gal})}{4{,}164 \text{ gpm}} = 240 \text{ min or 4 } \textit{hours}$$

SOR would be (2 basins × [37.75 × 120 ft] = 9,060 ft²,
and 4,164 gpm/9,060 ≈ *0.46 gpm/ft²*

Flow-through velocity would be (9.3 ft³/s × [60 s/min])
(2 × [37.75 × 14.8 ft]) ≈ *0.5 ft/min*

WOR = 6,000,000 gal/d/(48 ft × 2 sides × 3 weirs)(2 basins) =
10,416 gal/d/ft or *7.23 gpm/ft*

These values were approvable at the time but were rendered somewhat meaningless when tubes were put into the basins (filters were retrofitted to higher rate, as well). For example, taking a basin out of service for cleaning and maintenance in the days before tubes were installed meant that 50% of the rated flow, or 3 mgd, was unavailable for treatment. When the tubes were put in, the newly approved SOR at 9 mgd became 0.2 gpm/ft^2. This is a decrease of (0.46/0.2) = 2.3 times less the former aggressive rate and allows the superintendent to remove a basin from service with more confidence that the remaining one will perform well. She can now easily schedule maintenance and cleaning operations. Essentially, the plant got a 50% increase in treatment rate (from 6–9 mgd) and managed to gain a 230% improvement to the overflow rate, all in the same footprint.

Process question: Filter aid polymer—a sed basin was out of service for a short period for cleaning, and the plant manager decided to use some filter aid polymer at the end of the other basin just as a precaution. The operating staff makes up a 1% solution of polymer and wants to dose the water going to the filters at 0.05 mg/L. How many mL/min should they feed?

Answer: The operator uses the old standby formula of $V_1N_1 = V_2N_2$ and sets it up this way:

(4,164 gpm) × (0.05 mg/L) = (X gpm) × (10,000 mg/L), therefore:

$$\text{X gpm} = \left(4{,}164 \times 0.05 \big/ 10{,}000\right) = 0.02\text{-gpm filter aid needed}$$

(0.02 gpm × 3,785 mL/gal) = *75.7 mL/min needed*

Process question: Filters—given the following information on the filters at the WTP, calculate the unit filter run volume (UFRV) for filter #4, the L/De ratio of the design filter data, and the d_{90}/d_{10} ratio as designed.

Filter design data:

- Filters: number of units = 4; approved filtration rate = 5.7 gpm/ft^2
- Dimensions: 11 ft × 33 ft each; surface area per filter = 363 ft^2
- Filter capacity (each): approximately 3 mgd
- Capacity: one filter out of service = 9 mgd
- Media configuration: dual media
- Anthracite: 24 in., effective size (ES) = 1.11 mm; d_{60} = 1.43; d_{90} = 1.62 mm
- Sand: 12 in, 0.52-mm ES; d_{60} = 0.65 (except filter #3 at 0.83)

Filter production data for the month:

From the table, we see that the filter production was 10 runs each filter.

- Filter #1: 46,540,000 gal
- Filter #2: 44,700,000 gal

- Filter #3: 46,240,000 gal
- Filter #4: not available—totalizer O.O.S.

Answer: To calculate UFRV for filter #4, we first must add up the production values of the other three filters. Adding the production of filters #1–3, we get 137.48 MG for the month. That means that filter #4 must have produced 49.42 MG (186.9–137.48 MG).

Because the filters had 10 runs each, it is implied that filter #4 would have averaged (49.42 MG/10 runs) = 4.942 MG per run.

UFRV, according to the Ten State Standards, requires that a newly designed filter must demonstrate that it can produce at least 5,000 UFRVs before it can be approved (see Chapter 7).

UFRV is calculated by dividing the total water produced in the run by the total square feet of the filter, so 4,942,000 gal/363 ft^2 = *13,614 UFRV*, which is a good value to keep track of because it can help spot trends in the health of the filters. All of the filters should produce similar values from month to month. If one or more filters begin to deviate from their traditional UFRV value, they should be checked as soon as possible.

Notice that UFRV has no time element to it. Time is irrelevant in the case of filter performance. If someone tells you that their filters get 80 hours of run time, it doesn't tell you anything about how much water was produced relative to any of the other filters.

L/De ratio and d_{90}/d_{10} ratio are design parameters that engineers use to create effective dual-media filters. In the case of dual-media units composed of anthracite and sand, it is customary to build a filter with enough media so that the depth of it, converted to millimeters and divided by the ES, would equal a value of greater than 1,000. We can calculate the L/De for these filters as follows:

$$(\text{Anthracite} = 24 \text{ in.} \times 25.4 \text{ mm/in.})/1.11 \text{ mm} = 508$$

$$(\text{Sand} = 12 \text{ in.} \times 25.4 \text{ mm/in.})/0.52 = 586$$

$$L/De \text{ ratio} = 508 + 586 = 1{,}094$$

The d_{90}/d_{10} ratio tells us a lot about the zone that occurs at the sand/anthracite interface. A dual-media filter should have a zone that is not an abrupt change from one to the other, nor should there be too much mixing of the two layers. To accomplish that, the designers of the bed must choose filter media very carefully such that the largest of the anthracite particles (the D_{90} size) sits atop the smallest (10% size, or ES) of the sand particles. In technical terms, the D_{90} anthracite particle diameter should be somewhere between 3 and 3.5 times larger than the ES diameter of the sand particles (if the filter is granular activated carbon over sand, that ratio will be closer to 4.0). The calculation for these filters is as follows:

$$d_{90}/D_{10} \text{ ratio} = 1.62 \text{ mm}/0.52 \text{ mm} = 3.11$$

Process question: What might the superintendent expect of filter #3?

Answer: The uniformity coefficients of the sand in filters #1, #2, and #4 are all (0.65/0.52) = 1.25, but #3 has a uniformity coefficient for the sand of (0.83/0.52) = 1.6. It was probably installed incorrectly, or the sand may have come from a different batch from the vendor. It is not as uniform; it has a wider range of small and larger grain sizes mixed together and may not perform like the other filters do. It may produce lower overall head loss, but at the expense of higher effluent turbidities.

Process question: You are the operator on shift one ordinary day, when at 9:00 a.m., the filters begin to experience a sharp increase in loss of head. Refusing to panic, you inform the superintendent, who orders you to begin recycling that water so that all four filters can be washed. It is at that point that you notice that the SFBW tank is full. Sticking to Partnership for Safe Water performance goals, and cursing the midnight operator, you must calculate how long it will be before you can safely wash the first filter. For plant flow rate, use the average daily treated water flow. It takes 10 min to backwash a filter. At the high rate, you are washing at 20 gpm/ft^2.

Answer: You need to recycle the SFBW as soon as possible to empty the holding tank enough to fit the volume of one whole wash back into it, but you can't exceed the PSW 10% limit. From the data table and process description, the following applies:

Plant average treated water flow rate = 186,900,000 gal/month ÷ 1 month/30 d = 6,230,000 gal/d, or about 4,236 gpm. Therefore, you can recycle at a rate of (10% × 4,236) = 424 gpm.

Backwash rate for those 10-min averages: (1,785,960 gal wash water used/40 washes) ÷ 10 min = (44,641/10) = about 4,640 gpm average (of course, you'll start out slowly then hit a peak, and gradually back down). At peak backwash flow of 20 gpm/ft^2, you would be flowing (20 × 363) = 7,260 gpm into the holding tank. Thoughts of a quick series of back-to-back-to-back-to-back backwashes are out of the question.

The SFBW holding tank is full at 250,000 gal. You begin taking water out of it at a rate of 424 gpm. Keeping it safe and simple, you calculate a detention time problem:

$$\text{dt} = \text{Volume/Flow Rate} = (44{,}641 \text{ gal})$$
$$\text{for 1 Wash/}(4{,}640\text{–}464)\text{gpm} = 10 \text{ or so min}$$

It is best to wait a little over 10 min before you begin washing. Use the time to enroll in a calculus class at the local college.

Process question: Disinfection—what would the $C \times T$ (concentration times time) value be on a day when the clearwell was at its lowest point of 9.3 ft, the plant flow was 5,760 gpm, and the free chlorine residual through the clearwell was 1.7 mg/L?

Answer: The $C \times T$ value must be calculated and recorded, then reported for the worst-case conditions each day of the reporting month. Generally, the superintendent will have come to an agreement with the regulator on which flow rate to use because things can be quite complex if there are intermediate storage basins and relay pumps or if the disinfection chamber is a complex set of odd-sized basins in both series and parallel.

In this case, there is a simple calculation to perform:

$$C \times T = \left(\left(\frac{(2)\times\left(5{,}184\ \text{ft}^2 \times 9.3\ \text{ft} \times 7.47\ \text{gal/ft}^3\right)}{5{,}760\ \text{gpm}}\right\} \times 0.38\right) =$$

47 min of effective time (T10 time)

47 min × 1.7 mg/L = *80.8 mg-min/L*

Process question: Dosages—Calculate the dosages in mg/L for the following: dry-basis alum, post Cl_2, F^-, PO_4^-, $KMnO_4$, and powdered activated carbon (PAC).

Answer:

Dry-basis alum: At a specific gravity of 1.325, the liquid alum weighs (8.34 × 1.325) = 11 lb/gal. At 48.8%, there would be about (0.488 × 11.05) = 5.36 lb/gal as dry-basis alum. Treated water is 186.9 MG.

At 5,318.67 gal/month × 1 mo/186.9 MG = 28.6 gal/MG or ppm

28.6 gal/MG × (5.36 lb dry/gal) ÷ 8.34 = *18.3 mg/L*

As a check, we can have some fun and string it out:

5.36 lb dry-basis/gal × (453,600 mg/lb) = 2,431,296 mg/gal

5,318.6 gal fed/month × 2,431,296 mg/gal =
12,931,090,905.6 mg fed in month

12,931,090,905.6 mg/([186,900,000 gal] × 3.785 L/gal)) = *18.3 mg/L*

Post chlorine gas: The notes say that 1 mg/L was fed before the filters, so we must assume the rest was fed after filtration. To determine the dose, we set up our equation this way:

Dose in mg/L After Filtration = Total Dose, mg/L–1 mg/L

B4 dose = 1 mg/L × 8.34 lb/MG × 186.9 MG of
water treated/month = 1,560 lb of Cl_2

5,473 lb/month–1,560/month = 3,913 lb into
finished water pumped, therefore:

Post Cl_2 dose = 3,913 lb/month/30 d/5.92 MG/8.34 = *2.65 mg/L*

Process question: Fluoride ion dose—Because the feed point is before filtration, the WTP loses 0.22 mg/L of the F ion in the water at that point, but the source water has 0.14 mg/L natural fluoride in it. What might we expect the average F ion to be in the finished water (that is, during the times when wash water is not being recycled)? (Yes, all this actually happened at a plant I once supervised. The designer and the regulator both missed it in the approval process.)

Answer: Hydrofluosilicic acid is shown as H_2SiF_6, so the fluoride ion (F ion) content of the molecule itself is calculated by adding up the molecular weights of all the elements and subtracting the non-F constituents:

F = 6 × 19 = 114; H = 2 × 1 = 2; Si = 1 × 28 = 28; total is 144, and therefore, 114/144 = 79% available fluoride in the molecule of hydrofluosilicic acid.

The acid was listed as 23% strength and 10.5 lb/gal.

Using the source water flow rate of 186.9 MG and the gallons of acid as 654 weighing 10.5 lb/gal, calculated F ion is added as follows:

Product, mg/L = (789 gal × 10.5)/186.9 MG/8.34 =
5.31 mg/L of product

F ion, mg/L = 5.31 mg/L × 0.23 acid strength × 0.79 F^- in molecule =
0.96 mg/L fed in the month on average, but

0.96 mg/L Fed + 0.14 in Source Water–0.22 Loss = *0.88-mg/L F ion*

Phosphate, PO_4—Phosphoric acid was fed into the finished water to keep the lead and copper program in good shape. That finished water pumpage is listed as 5.92 mgd. During the month, the operators used 910 gal, or (910/30) = 30.3 gal per day of the phosphoric acid:

(30.3 gal × 12.2 lb/gal)/5.92/8.34 ≈ 7.5 mg/L product

7.5 mg/L × 0.30 ≈ 2.2 mg/L phosphoric acid

Phosphoric acid is H_3PO_4, so the molecular weight is H = 3 × 1; P = 1 × 31; O = 4 × 16, and then total weight is 98. PO_4 then is (95/98) = 97%, and 97% of 2.2 mg/L is about *2.1-mg/L phosphate.*

If the regulator wants the value reported as P, it would be about 0.33 times the phosphate, or about 0.7 mg/L.

Potassium permanganate, $KMnO_4$—The information given is that this product is made up in the day tank at a concentration of 0.2 lb/gal. The operators used 367.6 lb in the month; therefore, the concentration of the liquid contents of the day tank is not important for simple dosage calculations.

mg/L = 467.6 lb/186.9 MG/8.34 = *0.3 mg/L* $KMnO_4$

Activated carbon—This is also straightforward, although they don't tell us if it was used each day or only for a short period of a few days. Therefore, we are forced to calculate an average for the month using the source water flow, and it becomes the following:

PAC dose, mg/L = 4,475 lb/ 186.9/8.34 = *2.87-mg/L PAC*

Process question: pH adjustment—Normally, Lake Erie WTPs have little reason to need a pH adjustment chemical such as caustic soda (NaOH) under normal circumstances. The available alkalinity in the lake source water helps to maintain a higher pH in spite of the liquid alum products used for coagulation. However, as mentioned in earlier chapters, a dimictic lake such as Erie can develop a dead zone that has a very low pH, and forward-thinking utilities keep a small amount of NaOH on-hand so that pH can be adjusted upward during those three or four days of upheaval. This chemical comes in drums that carry 50% solution and helps them maintain the integrity of their painstakingly applied phosphate program for lead and copper minimization and avoid a secondary maximum contaminant level violation for manganese. The product weighs 12.76 ppg and comes in 55-gal drums that sit on a scale for "weight loss over time" measurements when making rounds.

The operator, who is savvy enough to not take supervisory control and data acquisition readings as gospel and doesn't like to guess and or be surprised at a time most inconvenient, decides to calculate the amount of caustic soda product that should be left in the drum in his 6 a.m. round. At the 2 a.m. round, he notes that there were 456.3 lb of product left in the drum. He's been treating a steady flow rate of 6 mgd with a dose of 3.5 mg/L NaOH. What should that be?

Answer: First, he uses the familiar dosage formula, and calculates that 3.5-mg/L NaOH × 8.34 lb/MG in 1 mg/L × 6 mgd = 175.14 ppd. And 175.14 ÷ 1,440 mi/d = 0.1216-lb/min NaOH.

0.1216 lb/min × 240 min = about 29 lb fed.

But he isn't using pure NaOH. Rather, it's 50% caustic:

29 lb NaOH/0.5 = about 58 lb should have been used by 6 a.m.

So, the drum should weigh about (456–58 lb) = *about 398 lb.*

RANDOM LOGIC QUESTIONS

The remainder of the problems that are included here are random logic (RL) problems that occur from time to time in WTPs and require some imagination by the operator. They don't pertain to or depend on the drawing and data previously shown. They are displayed as "multiple choice" with three or four answers to choose from; the answers are at the end of the chapter.

RL #1

This first one asks to calculate the pound per million gallons of chlorine needed to produce 0.4-mg/liter free chlorine residual.

Answers: A) 4.16 lb/MG B) 8.34 lb/MG C) 16.68 lb/MG

	A	B	C	D	E	F	G	H	I	J	K	L
Applied	0.2	0.4	0.6	0.8	1.0	1.2	1.4	1.6	1.8	2.0	2.2	2.4
Residual	0.1	0.2	0.3	0.4	0.5	0.6	0.4	0.1	0.2	0.4	0.6	0.8

RL #2

The accuracy of a flowmeter and totalizer for the pump and piping system that leads to the clearwell is being tested. The following information is provided:

1) The circular clearwell is 20 ft deep and has a diameter of 72 ft and is filled using a relay pump.
2) The water level in the clearwell is allowed to drop 36 in. to make some space for the test.
3) All flow is pumped into the clearwell through this line and Venturi meter. The beginning totalizer reading was 671,808.
4) The relay pump is turned on, and the water level rises 24 in. in 10 min.
5) The meter measures a flow rate of 6,697 gpm. The final totalizer reading was 732,687.

Determine if the totalizer is 10% high, accurate, or 9% low.
Determine if the flowmeter is 10% high, accurate, or 10% low.

RL #3

The following information is provided for a clearwell arrangement.

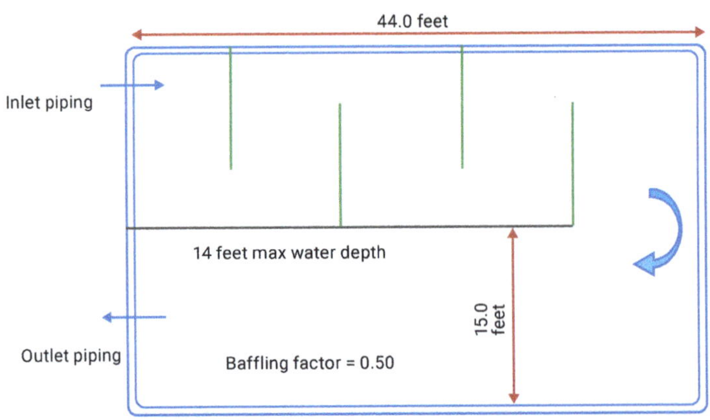

PLANT OPERATING DATA FOR CT EXAMPLE

Time	Depth *feet*	Cl_2 Residual *mm/L*	Flow *gpm*	Time	Depth *feet*	Cl_2 Residual *mm/L*	Flow *gpm*
00:00	11.0	1.5	220	12:00	9.6	1.5	175
01:00	11.0	1.5	220	13:00	9.0	1.5	175
02:00	10.9	1.5	220	14:00	8.7	1.5	175
03:00	10.9	1.5	220	15:00	8.5	1.5	220
04:00	10.6	1.4	220	16:00	8.4	1.6	220
05:00	10.5	1.5	220	17:00	8.3	1.6	220
06:00	10.0	1.5	280	18:00	8.3	1.6	300
07:00	9.8	1.7	270	19:00	8.6	1.6	280
08:00	9.7	1.7	270	20:00	9.0	1.5	280
09:00	9.7	1.7	270	21:00	9.2	1.5	220
10:00	9.9	1.6	220	22:00	9.5	1.5	220
11:00	10.0	1.6	230	23:00	9.9	1.5	220

Find the worst-case scenario: Which hour had the lowest disinfection sufficiency? Calculate that contact time (CT) value.

RL #4

The plant operator at an older facility has only one storage tank for sodium hypochlorite (NaOCl) in which she is storing 500 gal of a 4.5% solution. She has enough room to accept a new shipment of 800 gal of sodium hypochlorite at 12%. What strength can she expect the resultant 1,300-gal mixture to be?

A) 13% B) 7.5% C) 9.1% D) 10.8%

RL #5

Given the following information, calculate how many pounds per day of carbonates are deposited on the filter if the WTP operates at 5,560 gpm for 12 hours a day. The recarbonated water total and phenolphthalein alkalinity were 60 ft and 28 mg/L, and the filtered water was 40 and 6, respectively.

A) 734 lb/d B) 780 lb/d

B) 866 lb/d D) 701 lb/d

RL #6

You have 125 mL of sulfuric acid—H_2SO_4—in the lab, and you need to make 0.02*N* acid. You use 25 mL of the acid to neutralize 52 mL of 0.045*N* $NaHCO_3$. Calculate the amount of distilled or lab grade water needed to bring the remaining acid to the desired normality.

A) 468 mL

B) 368 mL

C) 268 mL

D) 168 mL

RL #7

Your WTP has an upground reservoir that can hold as much as seven days of source water, giving you the ability to pump water from the river into it when river water is of good quality and to avoid pulling form the river after runoff events. The watershed for the river is surrounded by farms and cattle feedlots. What lab tests should you run to determine when to pump water from the river into your storage?

A) NO_3^-, Cl^-

B) SO_4^-, PO_4^-, Cl^-

C) D.O., NO_3^-, SO_4^-

D) Algae, Cl^-, coliforms

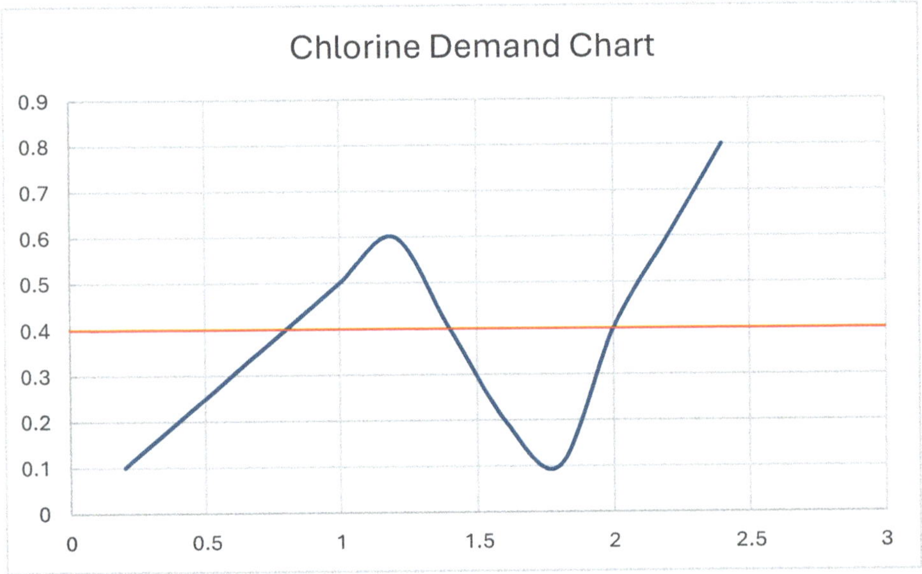

Answers to RL Questions

Answer #1

If we graph the dose versus the residual, it's evident that 0.4 mg/L residual occurs three times, but we have to find the 0.4 residual that corresponds to the dosage that results in free available Cl_2. That only occurs after the demand has been met, which is to the right after mono- and di-chloramines have been minimized. At 2 mg/L applied, it would be 2 mg/L × (8.34 lb/MG/1 mg/L) = 16.68 lb/MG.

Answer #2

This is admittedly the old-fashioned way of doing things, but it was effective in its time and operators swore by it. We know for a fact that the pump put 2 ft of water the clearwell in 10 min.

Ten feet of water is 0.785 × 2 ft × 72 ft × 72 ft × 7,48 = 60,879 gal.
That's 6.088 gpm.

The totalizer is telling us that it registered a total volume of water equal to (732,687 – 671,808) = 60,879 gal. That matches what we calculated, so the totalizer is accurate.

The flowmeter was registering flow rate of 6,697 gpm, but we know we pumped at a rate of 6.088 gpm for our 2-ft estimate. Therefore, the meter is (6,088 ÷ 6,679) = 9% high.

Answer #3

The lowest level of the clearwell was 8.3 ft, which occurred at 1,700 and 1,800. But the flow rate was its highest at 1,800 when it was at 300 gpm.

The volume of the clearwell at 1,800 would be calculated as follows:

Volume = 44 ft × 30 ft × 8.3 ft × 7.48 = 82,346 gal approximately

$$\text{CT value then is } \frac{82{,}346 \text{ gal} \times 0.5 \text{ baffling fact}}{300 \text{ gpm}} \times 2.6 \text{ mg/L } Cl_2 = 356.8 \text{ mg-min/L}$$

Answer #4

The procedure is to:

A) Multiply the amount and strength of each solution.

B) Add the two results together.

C) Divide that by the total of the two amounts you began with.

Example: What is the resultant solution strength of mixing 500 lb of 4.5% solution with 800 lb of 12% solution?

Answer: $\dfrac{(500 \times 0.045) + (800 \times 0.12)}{(500 + 800)}$

You will have 1,300 gal of 0.091, or 9.1% solution, or 91,000 mg/L.

Answer #5

This question simply requires us to look at the recarbonation process to see how efficient we are at it because the plant is not using a sequestering agent before filtration. What recarbonation does not remove, filters will need to handle.

If it's a matter of determining the amount of carbonate (CO_3^{-2}) before and after filtration to see how much carbonate loss is experienced, we can use the alkalinity table we saw in Chapter 9.

In the water before and after, the amount of P alkalinity is less than half of the T alkalinity, so carbonates will be 2P.

Carbonated: 2(28) = 56 mg/L

Filtered: 2(6) = 12 mg/L

Loss across the filter is 56–12 = 44 mg/L.

If the plant operated at 5,560 ft for 12 hours each day, they treated 5,560 gpm × 720 min/d = about 4 MG.

Therefore, lb/d = 44 mg/L × 8.34 × 4 MG = 1,468 lb/d

Answer 6

For this, we can start off with the ubiquitous formula that tells us that if the volume and concentration of one solution is known, then we can find the volume or the concentration of the second solution as long as at least one of those is known:

$$V_1N_1 = V_2N_2$$

What we want to learn is the strength of the first solution:

$N_1 = (V_2N_2) \div N_2$, or $N_1 = (52\text{ mL} \times 0.045) \div 25 = 0.0936N$, which is the strength of the original acid. Now, you've only got 100 mL of it left. So, how much water do you need?

$(100\text{ mL} \times 0.0936N) = V_2 \times 0.02N$, so $V_2 = 468$ mL

You would need to add (468 desired outcome–100 mL of original) = 368 mL of Water.

Answer 7

A river or stream subject to feed lot runoff and farming operations runoff will get polluted by cattle urine, fertilizers, and manure, and will lose its oxygen (O) content as the bacteria eat it up.

But because you can store only seven days of water, you have to make quick decisions on whether to pump water into your reservoir. So, you can't wait for coliform analysis. Nitrates will account for increases in fertilizer and manure, and chlorides will tell you if urine is present. Dissolved oxygen (DO) is an indication of healthy water, so the answer with the best choices would be A—nitrates, chlorides, and DO.

SUMMARY

The problems shown here are real-life situations that crop up in most WTPs. Hopefully, all of those who worked their way through the math got some useful information by doing so. Plant managers and operators can be so busy at times that they can't set aside time to work practice exercises periodically to ensure preparedness for the real thing.

Practicing the logical steps found in this chapter is one way to keep yourself out in front. If you can adapt these exercises to your own plant, all the better. It will help you and future generations of operators to keep the plant in compliance.

Chapter 17

Responsibility and Accountability

Responsibility refers to the obligation to perform a task or comply with the rules. Responsibility can be assigned to one person or a team of people.

Accountability implies answerability for the outcome of that task or compliance with those rules. Accountability is what we get if we combine responsibility with answerability. Accountability means the state of being responsible or answerable for a system, its behavior, and its potential effects. To be accountable is to acknowledge the responsibility for your actions, decisions, and the quality of your product.

OPERATOR ETHICS AND ACCOUNTABILITY

This book is filled with scientific facts, engineering basics, mathematical equations, and experienced-based operating knowledge that is accepted as best practice. This chapter, however, addresses the topic of ethics and operator culture that is based on morals and human nature. It is impossible for me to address this topic without offering an opinion based on my own experiences and beliefs. I do so, drawing from my 52 years in the business.

THE PURPOSE OF WATER SUPPLY

The reason why society has developed and regulated drinking water treatment systems is so that the community can rely on an uninterrupted source of safe water for all their needs. We can debate the priority of importance of those needs, but among them would be overall community health, firefighting capability, commercial and industrial growth capability, recreational activity, farming activities, and other essential uses.

THE INDIVIDUALS RESPONSIBLE

In an ideal world, the individuals who take employment in the water systems of this country would be ethical, educable, capable people who understand

and embrace the purpose of water supply and who are willing to pursue the achievement of that purpose at all times.

We do not live in an ideal world.

Operator Responsibility

The task of operating in a drinking water treatment plant (WTP) starts out for most individuals as an employment opportunity with an unknown future. No one is born with the innate sense of the tremendous responsibility or opportunity to serve the community that is attached to the local treatment works. To many, it is a means of making a living—a reliable and respectable way to pay the bills and put food on the table.

To a few, it becomes a career. The challenges, the opportunity to improve, the sense of purpose, and the need to provide a meaningful contribution to society—these take hold of some operators who, in time, cannot imagine doing anything else. It is not all fun and games, however.

The licensed "operator in charge" (OIC), or operator of record of the WTP, is one of the most challenging professions I've ever dealt with, and for most people who have had the experience, they would agree. Why is this so?

The licensed OIC is accountable to existential "groups of interest," some of whom have conflicting ideas about the way the OIC ought to perform. For many of these OICs, the day will come when one of these groups will expect them to perform contrary to each other. What that OIC does in that situation can make or break a career.

The OIC should be prepared for this possibility. If it comes, more than likely it won't be at an ideal time. That OIC may be young and raising a family, trying to rise up in the organization, or just trying to set an example for the staff. To help prepare for that day, the following discussion is offered.

Let's start by assuming that you are a recently promoted OIC with your career in front of you. You are in compliance with regulations, but you know the plant can improve in many ways by optimizing the efforts of plant staff and your processes. Who are your "groups of interest"?

- There is your employer and your immediate supervisor—the people who control your paycheck. Many OICs find themselves in a position of answering to a supervisor who may have goals for the department that are in conflict with water quality goals that have been set or that make compliance difficult. That supervisor may direct you to do something that is immoral or illegal. The reasons behind their goals or directives can come from a variety of sources.
  - Ignorance: They may not be aware of the implications of poor water quality or the harm it can inflict on a customer.
  - Egotism: They may think so highly of themselves that they cannot tolerate your intelligence or ambition, nor your efforts to "weigh in" or otherwise contradict their judgment.
  - Resentment: They may have deep-rooted abhorrence to governmental control of water systems. Their personal beliefs—indeed, even

the company culture itself—may resent the US Environmental Protection Agency and its regulatory requirements such as the Operator Certification Rule, which states that the water system be under the care of a licensed operator. They may feel that they are being forced to pay you a premium wage and that you are a necessary evil. What does not come under their control may be held against you.
  - Poor self-esteem: They may think so poorly of themselves that they see you as someone who will expose their limitations.
  - Indifference: They may not care about improved water quality. Perhaps they are an elected official who wants to be reelected or appointed, and your needs for the plant are untimely. Or they may be a manager in a private utility who has control of the budget and gets to keep a percentage of the money he saves as his bonus each year.
  - Fearfulness: They might be wary of upsetting the status quo by allowing you to implement new ideas.
- There is the regulator—the agency that owns the license that was issued to you and the body that can revoke that license. When you sign off at the end of each month, you are certifying that the data you send are truthful and that you followed regulated protocols, which is a legal issue. It is also a moral issue. When water quality doesn't meet standards, you are the person who gets called by the regulator, not your boss. And the regulator can't be bothered with excuses about your boss.
- There is the public—the people who expect you to keep learning and who depend on you to make the proper decisions that protect their health. (You might include the local press and other media as part of the public). All of this is expected of you by a public who thinks that water is either too expensive, tastes funny, or can't understand why you just don't give it away in buckets. The public often does not notice the local water department until something bad happens, such as a main break or a taste-and-odor episode or a boil notice.
- There is the staff—the group of individuals (yourself included) that have been assembled to operate the treatment facilities. Running a water system is a business, and it takes manpower to do it adequately. The individuals working in the plant are human beings who are not perfect. Among them will be individuals who are capable and willing to carry out their responsibilities, those who are not, and those who are dishonest or worse. The intellectual capabilities may vary from person to person, and the physical attributes may also. In this regard, the water business is no different from others; it attracts its share of malcontented employees. Working with and depending on them will be a challenge.

Water systems that are publicly owned and that come under the jurisdiction of the governing body (city council, commissioners, mayor, etc.) are sometimes subject to nepotism, have such low wages that they cannot attract the more qualified workers, or have aging facilities that discourage young people from taking part in what might otherwise become a successful adventure. Civil service laws of the state may make it impossible to discipline staff to any degree when warranted, which can discourage some operators. When they witness fellow operators who are nonperformers in their duties but are paid the same, they can become discouraged.

Privately owned systems may pay well but may be under the control of a board of directors who think of the utility as a source of profit. If the company management won't hesitate to cut corners whenever they see fit, and if you work directly for a person who has no experience in plant operations, it will be a challenge to get them to understand your needs. No company was ever formed for the sole purpose of losing money for its investors. But when your manager tells you to cut your chemical budget to the point of poorer water quality or when the WTP needs treatment processes but has been refused them because of budgetary constraints levied by individuals who know little about WTP operations, your resolve will be tested. When you witness outright cheating or tampering with data, you must act. It's your license that is at stake.

Human beings are fallible and driven by all sorts of stimuli. Never underestimate the power of ambition, avarice, jealousy, and negativity in anyone you work for or with. Realize that human beings get sick, get distracted, become discouraged, lack sleep, can't always learn, become angry, and sometimes just don't care. Understand that its human nature for many workers to bring their politics and religion to work with them, making it impossible for them to act without the baggage they carry. Be especially aware if any of this applies to yourself.

Life Experiences

Perhaps most importantly, do not think for one moment that anyone who has come to work with you in this most important labor of love is incapable of sabotaging the safety of the water under your care for their own purposes. I have seen it several times. I will share one of many incidents with you:

Early in my water utilities career as a lab chemist, I ran the daily wet chemistries and the bacteriological analyses from samples I collected from the source, the processes, and distribution. My duties also included daily inventory control and procurement of treatment chemicals, which involved keeping track of how much we were using each day and calculating when to order the next loads. I managed the dosage control reports and had to correlate them to the inventory. Each month, the two were supposed to match within reason. For example, if I calculated the alum dose each day in milligrams per liter based on the day tank fill and draw rates, they should support the monthly inventory report that showed how much was left in storage.

The third shift operator would fill the day tank each midnight and mark down how many gallons had been used and drawn from storage to fill the day tank.

One day, I noticed that our source water and filtered water alkalinities were about the same; there was little if any of the characteristic drop in alkalinity that would be expected in a WTP that used alum. Our daily plant tap sample showed higher than normal turbidity (we didn't use continuous reading turbidimeters back then), and so I brought these abnormalities to the plant manager's attention.

He questioned my record keeping and lab methodology. After he checked my math and my chemical results, the boss walked with me over to the chemical feed room and looked at the alum storage tank. I assured him that it was at a level it should be. I told him I checked it every day, and that we were feeding the amount of liquid we were supposed to be feeding.

The math and the visual evidence before us added up. I suggested we increase the feed rate. We did. We saw no improvement.

Rather, things kept worsening. A day and a half passed. We rechecked everything we could think of, including my records for the last batch of alum that had arrived weeks earlier. He had me make fresh reagents for my alkalinity tests and the pH standard curve. We borrowed a bottle of fresh formazin from another WTP to calibrate our turbidimeter. By now, word had spread to the staff and the main office that we were having difficulties. (For reference, this took place in 1973, when turbidity was more of an aesthetic quality of water than a health-related one.)

We went back to the feeder, and he peered into the day tank, thought for a moment, and told me to bring a sample bottle. He said, "Take a sample of the alum from the day tank and drive it to the alum provider in Cleveland and ask them to check the strength." I did. The strength was about one-fifth of what it should be. Unsure of what to do, I told the provider someone would be in touch. As inexperienced and naïve as I was, I kept wondering if we got a bad batch or if the alum had a brief life span to it. When I brought back the result to the boss, he got on the phone and made several calls.

Anyone who has filled a 100-gal day tank by gravity from a 5,000-gal storage tank that's full knows that the difference in elevation from storage to day tank allows for a quick fill rate of only a couple minutes. But when the storage tank level drops to the lowest levels, transfer of liquid from storage to day tank could take 15 min or more, time enough for the operator to grab a smoke or make a call or maybe nod off.

What does an operator do when he becomes distracted and allows a few hundred gallons of liquid alum to overflow the day tank and is lost down the sewer drain? Does he confess to the boss and help get things back in order?

That depends on the individual. Some would rather water down the alum in the storage tank than take the blame, hoping others won't figure it out. He did so, and he had me fooled but not the boss. Neither one of us knew back then to simply check the pH of the alum—but the boss understood human nature.

Frustrated, all I could think of was that I wanted to confront the guy and ask him if he thought it was funny watching me sweat for a couple of days while I second-guessed myself and my lab skills. But the boss told me to stay out of it and keep my nose clean and said he'd handle it. "Learn from this," he said. I did.

WHAT CAN YOU DO?

As operators, what should we do? How does an operator maintain integrity under these pressures? How do you remain calm when your instincts tell you otherwise? How can you become responsible and accountable to all these groups? How can you strive to maintain your commitment to your customers?

The answer isn't as simple as "it's your job—if you are passionate about it, you'll get it done." But you can work to educate your staff, your regulator, your employer, your public, and yourself. You can learn from experiences. The best source for help in these endeavors is AWWA. The association recognizes all these challenges and wants to help you with answers.

Attending AWWA district meetings at the local level will bring you into contact with others who toil under the same conditions as you do, and that can help you cope. Get involved as a volunteer on committees that suit your talents and interests. This will expose you to new ideas and to people who have experiences they are willing to share with you. Forming alliances with fellow plant managers in your region may someday pay off in times of need.

AWWA has developed courses to help educate all utility workers, including management, civic leaders and elected officials, and superiors who do not have a background in plant operations. The association can help with public opinion and how to work with the local media. Perhaps best of all, AWWA can provide you with the materials you need to take exams and grow as a professional.

Take advantage of any employee career advancement opportunities your employer may provide. They might pay for college courses, seminars, and training events or allow you to bring in outside contractors who specialize in training services.

As your knowledge base grows and you learn more about your tasks and what you can expect of yourself, no one can persuade you to do anything but what you know to be right. If your heart is in the right place and if you have truly found your calling, you have no choice but to be faithful to the cause. You will be the operator of record. You'll be accountable to yourself.

Appendix A

SAMPLE
SAFETY DATA SHEET

Issuing Date January 5, 2015 **Revision Date** June 12, 2015 **Revision Number** 1

1. IDENTIFICATION OF THE SUBSTANCE/PREPARATION AND OF THE COMPANY/UNDERTAKING

Product identifier

Product Name **XXXXX Regular-Bleach[1]**

Other means of identification

EPA Registration Number 5813-100

Recommended use of the chemical and restrictions on use

Recommended use Household disinfecting, sanitizing, and laundry bleach

Uses advised against No information available

Details of the supplier of the safety data sheet

Supplier Address
The XXXXX Company
1221 Broadway
Oakland, CA 94612

Phone: 1-510-XXX-XXXX

Emergency telephone number

Emergency Phone Numbers For Medical Emergencies, call: 1-800-446-1014
For Transportation Emergencies, call Chemtrec: 1-800-424-9300

Source: Occupational Safety and Health Administration

2. HAZARDS IDENTIFICATION

Classification

This chemical is considered hazardous by the 2012 OSHA Hazard Communication Standard (29 CFR 1910.1200).

Skin corrosion/irritation	Category 1
Serious eye damage/eye irritation	Category 1

GHS Label elements, including precautionary statements

Emergency Overview

Signal word **Danger**

Hazard Statements
Causes severe skin burns and eye damage
Causes serious eye damage

Appearance Clear, pale yellow **Physical State** Thin liquid **Odor** Bleach

Precautionary Statements - Prevention
Wash face, hands, and any exposed skin thoroughly after handling.
Wear protective gloves, protective clothing, face protection, and eye protection such as safety glasses.

Precautionary Statements - Response
Immediately call a poison center or doctor.
If swallowed: Rinse mouth. Do NOT induce vomiting.
If on skin (or hair): Take off immediately all contaminated clothing. Rinse skin with water. Wash contaminated clothing before reuse.
If inhaled: Remove person to fresh air and keep comfortable for breathing.
Specific treatment (see supplemental first aid instructions on this label).
If in eyes: Rinse cautiously with water for several minutes. Remove contact lenses, if present and easy to do. Continue rinsing.

Precautionary Statements - Storage
Store locked up.

Precautionary Statements - Disposal
Dispose of contents in accordance with all applicable federal, state, and local regulations.

Hazards not otherwise classified (HNOC)
Although not expected, heart conditions or chronic respiratory problems such as asthma, chronic bronchitis, or obstructive lung disease may be aggravated by exposure to high concentrations of vapor or mist.

Product contains a strong oxidizer. Always flush drains before and after use.

Unknown Toxicity
Not applicable.

Other information
Very toxic to aquatic life with long lasting effects.

Interactions with Other Chemicals
Reacts with other household chemicals such as toilet bowl cleaners, rust removers, acids, or products containing ammonia to produce hazardous irritating gases, such as chlorine and other chlorinated compounds.

Chemical Name	CAS-No	Weight %	Trade Secret
Sodium hypochlorite	7681-52-9	5–10	*

* The exact percentage (concentration) of composition has been withheld as a trade secret.

4. FIRST AID MEASURES

First aid measures

General Advice Call a poison control center or doctor immediately for treatment. Show this safety advice data sheet to the doctor in attendance.

Eye Contact Hold eye open and rinse slowly and gently with water for 15–20 minutes. Remove contact lenses, if present, after the first 5 minutes, then continue rinsing eye. Call a poison control center or doctor for treatment advice.

Skin Contact Take off contaminated clothing. Rinse skin immediately with plenty of water for 15–20 minutes. Call a poison control center or doctor for treatment advice.

Inhalation Move to fresh air. If breathing is affected, call a doctor.

Ingestion Have person sip a glassful of water if able to swallow. Do not induce vomiting unless told to do so by a poison control center or doctor. Do not give anything by mouth to an unconscious person. Call a poison control center or doctor immediately for treatment advice.

Protection of First-aiders Avoid contact with skin, eyes, and clothing. Use personal protective equipment as required. Wear personal protective clothing (see Section 8).

Most important symptoms and effects, both acute and delayed

Most Important Symptoms and Effects Burning of eyes and skin.

Indication of any immediate medical attention and special treatment needed

Notes to Physician Treat symptomatically. Probable mucosal damage may contraindicate the use of gastric lavage.

5. FIRE-FIGHTING MEASURES

Suitable Extinguishing Media
Use extinguishing measures that are appropriate to local circumstances and the surrounding environment.

Unsuitable Extinguishing Media
CAUTION: Use of water spray when fighting fire may be inefficient.

Specific Hazards Arising from the Chemical
This product causes burns to eyes, skin, and mucous membranes. Thermal decomposition can release sodium chlorate and irritating gases and vapors.

Explosion Data

Sensitivity to Mechanical Impact	None.
Sensitivity to Static Discharge	None.

Protective equipment and precautions for firefighters
As in any fire, wear self-contained breathing apparatus pressure-demand, MSHA/NIOSH (approved or equivalent) and full protective gear.

6. ACCIDENTAL RELEASE MEASURES

Personal precautions, protective equipment, and emergency procedures

Personal Precautions	Avoid contact with eyes, skin, and clothing. Ensure adequate ventilation. Use personal protective equipment as required. For spills of multiple products, responders should evaluate the MSDSs of the products for incompatibility with sodium hypochlorite. Breathing protection should be worn in enclosed and/or poorly-ventilated areas until hazard assessment is complete.
Other Information	Refer to protective measures listed in Sections 7 and 8.

Environmental precautions

Environmental Precautions	This product is toxic to fish, aquatic invertebrates, oysters, and shrimp. Do not allow product to enter storm drains, lakes, or streams. See Section 12 for ecological Information.

Methods and material for containment and cleaning up

Methods for Containment	Prevent further leakage or spillage if safe to do so.
Methods for Cleaning Up	Absorb and containerize. Wash residual down to sanitary sewer. Contact the sanitary treatment facility in advance to assure ability to process washed-down material.

Precautions for safe handling

Handling Handle in accordance with good industrial hygiene and safety practice. Avoid contact with skin, eyes, and clothing. Do not eat, drink, or smoke when using this product.

Conditions for safe storage, including any incompatibilities

Storage Store away from children. Reclose cap tightly after each use. Store this product upright in a cool, dry area, away from direct sunlight and heat to avoid deterioration. Do not contaminate food or feed by storage of this product.

Incompatible Products Toilet bowl cleaners, rust removers, acids, and products containing ammonia.

8. EXPOSURE CONTROLS/PERSONAL PROTECTION

Control parameters

Exposure Guidelines

Chemical Name	ACGIH TLV	OSHA PEL	NIOSH IDLH
Sodium hypochlorite 7681-52-9	None	None	None

ACGIH TLV: American Conference of Governmental Industrial Hygienists - Threshold Limit Value. OSHA PEL: Occupational Safety and Health Administration - Permissible Exposure Limits. NIOSH IDLH: Immediately Dangerous to Life or Health.

Appropriate engineering controls

Engineering Measures
Showers
Eyewash stations
Ventilation systems

Individual protection measures, such as personal protective equipment

Eye/Face Protection If splashes are likely to occur: Wear safety glasses with side shields (or goggles) or face shield.

Skin and Body Protection Wear rubber or neoprene gloves and protective clothing such as long-sleeved shirt.

Respiratory Protection If irritation is experienced, NIOSH/MSHA approved respiratory protection should be worn. Positive-pressure supplied air respirators may be required for high airborne contaminant concentrations. Respiratory protection must be provided in accordance with current local regulations.

Hygiene Measures Handle in accordance with good industrial hygiene and safety practice. Wash hands after direct contact. Do not wear product-contaminated clothing for prolonged periods. Remove and wash contaminated clothing before re-use. Do not eat, drink, or smoke when using this product.

9. PHYSICAL AND CHEMICAL PROPERTIES

Physical and Chemical Properties

Physical State	Thin liquid		
Appearance	Clear	**Odor**	Bleach
Color	Pale yellow	**Odor Threshold**	No information available

Property	**Values**	**Remarks/ Method**
pH	~12	None known
Melting/freezing point	No data available	None known
Boiling point / boiling range	No data available	None known
Flash Point	Not flammable	None known
Evaporation rate	No data available	None known
Flammability (solid, gas)	No data available	None known
Flammability Limits in Air		
Upper flammability limit	No data available	None known
Lower flammability limit	No data available	None known
Vapor pressure	No data available	None known
Vapor density	No data available	None known
Specific Gravity	~1.1	None known
Water Solubility	Soluble	None known
Solubility in other solvents	No data available	None known
Partition coefficient: n-octanol/ water	No data available	None known
Autoignition temperature	No data available	None known
Decomposition temperature	No data available	None known
Kinematic viscosity	No data available	None known
Dynamic viscosity	No data available	None known
Explosive Properties	Not explosive	
Oxidizing Properties	No data available	

Other Information

Softening Point	No data available
VOC Content (%)	No data available
Particle Size	No data available
Particle Size Distribution	No data available

10. STABILITY AND REACTIVITY

Reactivity
Reacts with other household chemicals such as toilet bowl cleaners, rust removers, acids, or products containing ammonia to produce hazardous irritating gases, such as chlorine and other chlorinated compounds.

Chemical stability
Stable under recommended storage conditions.

Possibility of Hazardous Reactions
None under normal processing.

Conditions to avoid
None known based on information supplied.

Incompatible materials
Toilet bowl cleaners, rust removers, acids, and products containing ammonia.

Hazardous Decomposition Products
None known based on information supplied.

11. TOXICOLOGICAL INFORMATION

Information on likely routes of exposure

Product Information

Inhalation	Exposure to vapor or mist may irritate respiratory tract and cause coughing. Inhalation of high concentrations may cause pulmonary edema.
Eye Contact	Corrosive. May cause severe damage to eyes.
Skin Contact	May cause severe irritation to skin. Prolonged contact may cause burns to skin.
Ingestion	Ingestion may cause burns to gastrointestinal tract and respiratory tract, nausea, vomiting, and diarrhea.

Component Information

Chemical Name	LD50 Oral	LD50 Dermal	LC50 Inhalation
Sodium hypochlorite 7681-52-9	8200 mg/kg (Rat)	>10,000 mg/kg (Rabbit)	-

Information on toxicological effects

Symptoms	May cause redness and tearing of the eyes. May cause burns to eyes. May cause redness or burns to skin. Inhalation may cause coughing.

Delayed and immediate effects as well as chronic effects from short and long-term exposure

Sensitization	No information available.
Mutagenic Effects	No information available.
Carcinogenicity	The table below indicates whether each agency has listed any ingredient as a carcinogen.

Chemical Name	ACGIH	IARC	NTP	OSHA
Sodium hypochlorite 7681-52-9	-	Group 3	-	-

IARC (International Agency for Research on Cancer)
Group 3 - Not Classifiable as to Carcinogenicity in Humans

Reproductive Toxicity	No information available.
STOT - single exposure	No information available.
STOT - repeated exposure	No information available.
Chronic Toxicity	Carcinogenic potential is unknown.
Target Organ Effects	Respiratory system, eyes, skin, gastrointestinal tract (GI).
Aspiration Hazard	No information available.

Numerical measures of toxicity - Product Information

The following values are calculated based on chapter 3.1 of the GHS document

ATEmix (oral)
54 g/kg
ATEmix (inhalation-dust/mist)
58 mg/L

12. ECOLOGICAL INFORMATION

Ecotoxicity
Very toxic to aquatic life with long lasting effects.

This product is toxic to fish, aquatic invertebrates, oysters, and shrimp. Do not allow product to enter storm drains, lakes, or streams.

Persistence and Degradability
No information available.

Bioaccumulation
No information available.

Other adverse effects
No information available.

13. DISPOSAL CONSIDERATIONS

Disposal methods
Dispose of in accordance with all applicable federal, state, and local regulations. Do not contaminate food or feed by disposal of this product.

Contaminated Packaging
Do not reuse empty containers. Dispose of in accordance with all applicable federal, state, and local regulations.

14. TRANSPORT INFORMATION

DOT	Not restricted.
TDG	Not restricted for road or rail.
ICAO	Not restricted, as per Special Provision A197, Environmentally Hazardous Substance exception.
IATA	Not restricted, as per Special Provision A197, Environmentally Hazardous Substance exception.
IMDG/IMO	Not restricted, as per IMDG Code 2.10.2.7, Marine Pollutant exception.

15. REGULATORY INFORMATION

Chemical Inventories

TSCA	All components of this product are either on the TSCA 8(b) Inventory or otherwise exempt from listing.
DSL/NDSL	All components are on the DSL or NDSL.

TSCA - United States Toxic Substances Control Act Section 8(b) Inventory
DSL/NDSL - Canadian Domestic Substances List/Non-Domestic Substances List

U.S. Federal Regulations

SARA 313
Section 313 of Title III of the Superfund Amendments and Reauthorization Act of 1986 (SARA). This product does not contain any chemicals which are subject to the reporting requirements of the Act and Title 40 of the Code of Federal Regulations, Part 372

SARA 311/312 Hazard Categories

Acute Health Hazard	Yes
Chronic Health Hazard	No
Fire Hazard	No
Sudden Release of Pressure Hazard	No
Reactive Hazard	No

Clean Water Act
This product contains the following substances which are regulated pollutants pursuant to the Clean Water Act (40 CFR 122.21 and 40 CFR 122.42)

Chemical Name	CWA - Reportable Quantities	CWA - Toxic Pollutants	CWA - Priority Pollutants	CWA - Hazardous Substances
Sodium hypochlorite 7681-52-9	100 lb			X

CERCLA
This material, as supplied, contains one or more substances regulated as a hazardous substance under the Comprehensive Environmental Response Compensation and Liability Act (CERCLA) (40 CFR 302)

Chemical Name	Hazardous Substances RQs	Extremely Hazardous Substances RQs	RQ
Sodium hypochlorite 7681-52-9	100 lb	-	RQ 100 lb final RQ RQ 45.4 kg final RQ

EPA Statement
This chemical is a pesticide product registered by the Environmental Protection Agency and is subject to certain labeling requirements under federal pesticide law. These requirements differ from the classification criteria and hazard information required for safety data sheets and for workplace labels of non-pesticide chemicals. Following is the hazard information as required on the pesticide label:

> **DANGER: CORROSIVE.** Causes irreversible eye damage and skin burns. Harmful if swallowed. Do not get in eyes, on skin, or on clothing. Wear protective eyewear and rubber gloves when handling this product. Wash thoroughly with soap and water after handling and before eating, drinking, chewing gum, using tobacco, or using the restroom. Avoid breathing vapors and use only in a well-ventilated area.

US State Regulations

California Proposition 65
This product does not contain any Proposition 65 chemicals.

U.S. State Right-to-Know Regulations

Chemical Name	New Jersey	Massachusetts	Pennsylvania	Rhode Island	Illinois
Sodium hypochlorite 7681-52-9	X	X	X	X	
Sodium chlorate 7775-09-9	X	X	X		

International Regulations

Canada
WHMIS Hazard Class
E - Corrosive material

16. OTHER INFORMATION

NFPA	**Health Hazard** 3	**Flammability** 0	**Instability** 0	**Physical and Chemical Hazards** -
HMIS	**Health Hazard** 3	**Flammability** 0	**Physical Hazard** 0	**Personal Protection** B

Prepared By Product Stewardship
23 British American Blvd.
Latham, NY 12110
1-800-572-6501

Revision Date June 12, 2015

Revision Note Revision Section 14.

Reference 1096036/164964.159

General Disclaimer
The information provided in this Safety Data Sheet is correct to the best of our knowledge, information, and belief at the date of its publication. The information given is designed only as a guidance for safe handling, use, processing, storage, transportation, disposal, and release and is not to be considered a warranty or quality specification. The information relates only to the specific material designated and may not be valid for such material used in combination with any other materials or in any process, unless specified in the text.

End of Safety Data Sheet

This material was produced under grant 46F6-HT30 from the Occupational Safety and Health Administration, U.S. Department of Labor. It does not necessarily reflect the views or policies of the U.S. Department of Labor, nor does mention of trade names, commercial products, or organizations imply endorsement by the U.S. Government. Revisions were made to this material under grant number SH-31240-SH7 from the Occupational Safety and Health Administration, U.S. Department of Labor.

Appendix B

Math and Calculation Methods

PRACTICAL UNIT CONVERSIONS FOR WATER

Cubic Feet (ft^3) to gallons (gal)

1 ft^3 of water is equivalent to 7.48 gal at normal temperatures.

Example: If a residential water meter registers in units of 100 ft^3, and 25 units of water have passed through the meter, how many gallons is this? 1 ft^3 is 7.48 gal, so $25 \times 100 \times 7.48$ gal/ft^3 = 18,700 gal.

Gallons (gal) of Water to Pounds (lb)

1 gal of water weighs approximately 8.34 lb.

Example: How many pounds of water are contained in a 50-gal day tank? $50 \times 8.34 = 417$ lb

Gallons (gal) to Liters (L)

1 gal is equivalent to 3.785 L, or 3,785 mL.

Example: An operator transfers 36 gal of polymer to a day tank. How many liters is this? 1 gal is 3.785 L, so $36 \times 3.785 = 136.26$ L; this is 136,260 mL.

Grams (g) to Pounds (lb)

1 lb is equivalent to 453.6 g.

Example: An operator feeds 13.5 lb of chemical each hour. How many grams is this? $453.6 \times 13.5 = 6{,}123.6$ g

Million Gallons per Day (mgd) to Cubic Feet per Second (ft^3/s)

1 mgd is equivalent to 1.55 ft^3/s.

Example: What is the average flow rate, in cubic feet per second, in a plant that produces 20 mgd? 1 mgd = 1.55 ft^3/s, so 20 mgd × 1.55 = 31 ft^3/s.

Million Gallons per Day (mgd) to Gallons per Minute (gpm)

1 mgd is equivalent to 694 gpm.

Example: If a water treatment plant (WTP) is operating at 6.75 mgd, how many gallons per minute is this? 1 mgd = 694 gpm, so 6.75 × 694 = 4,684.5 gpm.

Cubic Feet per Second (ft^3/s) to Gallons per Minute (gpm)

1 ft^3/s = 448.8 gpm.

Example: How many gallons per minute are represented in a flow of 3 ft^3/s? 3 ft^3/s × 60 s/min = 180 ft^3/min; 180 ft^3/min × 7.48 gal = 1346.4 gpm.

Percent (%) Concentration to Milligrams per Liter (mg/L)

1% is equivalent to 10,000 mg/L.

Example: An operator makes a 0.75% polymer solution in a day tank. How many milligrams per liter is this? 1% = 10,000 mg/L, so 0.75% × 10,000 = 7,500 mg/L.

Feet (ft) of Head to Pounds per Square Inch (psi) Pressure

1 psi is equivalent to a column of water that stands vertically at 2.31 ft.

Example: How many feet of head are developed at the base of a water tank that is 135 ft high? 2.31 ft of head = 1 psi, so 135 ft/2.31 = 58.4 psi.

Pounds per Square Inch (psi) Pressure to Feet (ft) of Head

1 ft of head is equivalent to 0.433 psi.

Example: How many feet of head are developed in a pumping system at 68 psi? 1 ft of head = 0.433 psi, so 68 psi/0.433 = 157 ft.

Grains per Gallon (gpg) to Milligrams per Liter (mg/L)

1 gpg = 17.1 mg/L.

Example: What is a water hardness of 11 gpg in mg/L? 11gpg × 17.1 mg/L for each gpg = 188.1 mg/L hardness.

Time Conversions

1 day is equivalent to 24 hours, which is equivalent to 1,440 min. 1 min is equivalent to 60 s. There are 86,400 s in 1 day.

PRACTICAL WTP EXAMPLE PROBLEMS

Volume

What is the volume, in gallons, of a rectangular clearwell that is 38 ft × 75 ft if the water depth is 14 ft? 38 ft × 75 ft × 14 ft × 7.48 gal/ft³ = 298,452 gal.

How many cubic feet of water are contained in 1,000 ft of 8-in. water main? Volume = $\pi R^2 H$ = 3.14 × 0.33 ft × 0.33 ft × 1,000 = 341.95 ft³.

What is the volume, in gallons, of a 60-ft-diameter tank when the water depth is 30 ft? Volume = 0.785 × (D^2) × depth × 7.48 gal/ft³ = 0.785 × (60 ft) (640 ft) × 30 ft × 7.48 = 634,154 gal.

NOTE: R is radius, and D is Diameter.

Solution Strength

An operator has a 10% stock solution of polymer and wishes to make 30 gal of a 1.5% working solution from the stock. How many gallons of the stock are needed? (30 gal)(15,000 mg/L) = (x gal)(100,000 mg/L); therefore, x = 4.5 gal.

If 500 lb of a 4.5% solution is mixed with 800 lb of a 12% solution, what is the resultant solution strength? (500 lb × 0.045) + (800 lb × 0.12)/(500 + 800) = 9.1%, or 91,000 mg/L.

Feed Rate

The operator makes the 1.5% working solution of polymer in the previous example. If the operator wants to treat a flow of 1,200 gpm with 0.5 mg/L of the polymer, how many milliliters per minute are needed? (1,200 gpm) (0.5 mg/L) = (x gpm)(15,000 mg/L); therefore, x = 0.04 gpm feed needed. And 0.04 gpm × 3.785 L/gal = 0.1514 L/min, or 151 mL/min.

An operator feeds activated carbon at a rate of 23 lb/h to treat a raw flow of 32 mgd. How many milligrams per liter is fed? 23 lb/h × 24 h = 552 lb/d; 552 lb/d/32 mgd = 17.25 lb/mil gal; 17.25 lb/mil gal/8.34 = 2 mg/L.

The gas chlorinators at a water plant go out of service; the operator decides to feed a hypochlorite solution that is 15% as available chlorine. If the operator needs to treat a flow of 1.6 mgd with 6 mg/L chlorine, how many gallons per hour hypochlorite solution are needed? 1,600,000 gpd/24 h = 67,000 gph; (67,000 gph × 6 mg/L)/150,000 mg/L = 2.68 gph.

At a WTP operating at 4 mgd, the dry alum feeder is set at an indicated feed rate of 17.3 mg/L. The operator runs a calibration check by catching 0.34474 kg of dry alum in 2 min. Is the dosimeter accurate? 0.34474 × 1,000 = 344.74 g; (344.74 g/2 min) × (1 lb/453.6 g) = 0.38 lb/min actually fed. The dosimeter is set at 17.3 mg/L, which is (17.3 × 8.34 × 4 mgd)/1,440 min = 0.40 lb/min. The dosimeter is not accurate.

Detention Time

Detention time (*DT*) = *V*/*Q*, where "*V*" is the volume in cubic feet and *Q* = the flow in cubic feet per second or per minute or per hour.

If a 4.52-mgd pump is used to fill a rectangular basin that is 75 ft × 36 ft, how long will it take to bring the water level from empty to 8 ft? *V* = 75 ft × 36 ft × 8 ft = 21,600 ft^3; *Q* = 4.52 mgd × 1.55 = 7 ft^3/sec; therefore, *DT* = 21,600 ft^3/7 ft^3/s = 3,085.7 s, or 51.4 min.

What is the carbon contact time if powdered activated carbon is fed into a 36-in.-diameter source water force main 23,000 ft in length and the flow through the force main is 8 mgd? 8,000,000 gal/d × 1 d/24 h × 1 ft^3/7.48 gal = 44,563 ft^3/h; *DT* = *V*/flow = (3.14 × 1.5 ft × 1.5 ft × 23,000 ft)/44,563 ft^3/h = 3.64 hours.

A wash-water recycle basin that measures 25 ft × 80 ft × 15 ft is emptied by a recycle pump operating at 560 gpm. Filters are being backwashed at the same time, and the flow from the backwash enters the recycle basin. Each filter backwash uses 54,600 gal. How many filters can be backwashed over a 4-hour period, assuming the basin was empty from the start? Basin volume is 25 ft × 80 ft × 15 ft × 7.48 = 224,000 gal. Basin is emptied at 560 gpm for 4 hours, which is 560 gpm × 60 min × 4 hours = 134,000 gal. Therefore, effective volume of the recycle basin is 134,000 gal + 224,000 gal = 358,000 gal; 358,000 gal/54,600 gal/wash = 6.57 washes. Six filters can be washed, but not seven.

Well Drawdown

The level of the aquifer of a well before pumping is 130 ft. When the pump is operated, the aquifer level drops to 173 ft. What is the drawdown? 173 ft – 130 ft = 43 ft.

Well Yield and Specific Yield of Wells

A total of 950 gal was pumped from a well in a period of 5 min. What is the well yield in gpm? 950 gal/5 min = 190 gpm.

Using the well yield in the previous example, calculate the specific yield if the drawdown is 25 ft. 190 gpm/25 ft = 7.6 gpm/ft.

Disinfection of System Components

An operator wishes to disinfect a new well casing that has a 12-in. diameter. The desired chlorine dosage is 50 mg/L. The casing is 150-ft long, and the water level in the casing is 55 ft from the top of the well. How many pounds of chlorine are needed? The water column is 150 ft – 55 ft = 95 ft. Therefore, the volume is 0.785 × 1 ft × 1 ft × 95 ft × 7.48 = 558 gal; Dosage = 50 mg/L × (558/1,000,000) × 8.34 = 0.23 lb of chlorine.

Determine the amount of 65% calcium hypochlorite needed to disinfect the water tower and riser in Figure B-1 if a 50-mg/L dose is desired.

Head

The static water level of a well pump is 90 ft. The well drawdown is 29 ft. If the gauge reading at the pump discharge head is 3.9 psi, what is the total pumping head? (90 ft + 29 ft) + (3.9 psi × 2.31) = 128 ft.

Using the information depicted in Figure B-2, determine the pressure at the main. Mercury is 13.6 times as heavy as water. The pressure will be the difference between the force created by the mercury and the force created by the water, so (3 × 13.6 ft) – (3 × 1 ft) = 37.8 ft of head. 37.8 ft × 0.433 = 16.4 psi.

Density and Specific Gravity

Density is usually expressed as weight per unit volume, such as pounds per gallon. Specific gravity is a ratio of the density of a liquid to the density of water. The density of water at room temperature is approximately 8.34 lb/gal. The density of water can vary with temperature; in fact, water is most dense

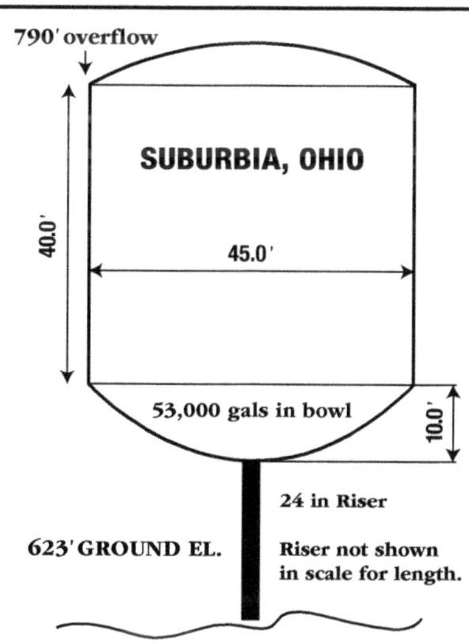

Tank is operated to overflow each night before plant is shut down. Tank flows on system during the night, and is refilled the next day.

Volume of tank = 3.14 × 22.5 ft × 22.5 ft × 40 ft × 7.48 gal/ft^3 = 475,615 gal

Volume of bowl = 53,000 gal

Volume of riser = 3.14 × 1 ft × 1 ft × (740 ft – 623 ft) × 7.48 gal/ft^3 = 2,748 gal

Total Volume = 475,615 + 53,000 + 2,748 = 531,363 gal, or 0.531 mg

Needed calcium hypochlorite = (0.521 mg × 50 mg/L × 8.34)/0.65 = 340 lb

Figure B-1 Disinfection of water tower

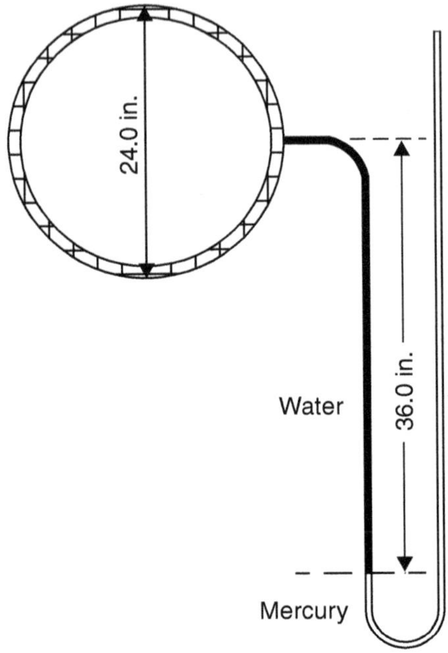

Figure B-2 Mercury manometer setup on 24-in. main

at 39.4°F (4°C). If 1 gal of oil weighs 7.51 lb, what is the specific gravity? (7.51 lb/gal)/(8.34 lb/gal) = 0.90.

The specific gravity of a batch of liquid alum is 1.295. What is the density? 1.295 × 8.34 lb/gal = 10.8 lb/gal. Each gallon of alum fed, in this case, would equate to 10.8 lb of liquid alum.

The specific gravity of mercury is 13.6. What does 1 gal of mercury weigh? 13.6 × 8.34 = 113.4 lb.

Flow Rate and Velocity

The flow rate calculation for water moving through a pipe or channel is $Q = A \times V$, where Q is flow rate in cubic feet per second, A is the square foot area of the pipe or channel, and V is the velocity in feet per second.

Newly installed water mains are normally flushed at 2.5 fps. How many cubic feet per second are needed to flush a 12-in.-diameter main? First, convert the 12-in. diameter to a radius in feet. 12 in./2 = 6-in. radius, and 6 in./12 = 0.5 ft; Q = 3.14 × 0.5 ft × 0.5 ft × 2.5 fps = 1.96 ft^3/s.

A water plant is operating at 3.6 mgd. After the flow travels through the rapid mix, it splits into two equally sized channels on its way to the flocculators. What is the velocity through one channel if it is 1.8 ft wide and the water depth is 18 in.? 3.6 mgd × 1.55 = 5.58 ft^3/s; 5.58 ft^3/s/2 = 2.79 ft^3/s through each channel; $V = Q/A$, or V = 2.79 ft^3/s/(1.8 ft × 1.5 ft) = 1 ft/s.

What is the velocity in meters per second in a pipe with a diameter of 42 cm and a flow of 7,500 m^3/d? 42-cm diameter = 0.42-m diameter, and 0.42/2 = 0.21-m radius. $V = Q/A$ = 7,500 $m^3/d/(3.14 \times 0.21 \text{ m} \times 0.21 \text{ m})$ = 54,161.8 m/d or 0.63 m/s.

The velocity in an 8-in. water main is 2 fps. How many gpm are flowing through it? Diameter of main is 8 in., so radius in feet is (8 in./1 ft/12 in.)/2 = 0.34-ft^2 area. Velocity is 2 ft/s × 60 s/min = 120 ft/min; $Q = A \times V$ = 0.34 ft^2 × 120 ft/min × 7.45 gal/ft^3 = 305 gpm.

Weir Overflow and Surface Overflow Rates

Weir overflow rates are expressed as gallons per day per foot, or gallons per minute per foot, while surface overflow rates are expressed as gallons per day per square foot or gallons per minute per square foot. Weir overflow rates use the total linear distance of the weirs at the end of the basins for a calculation, and surface overflow rates use the square foot area of the surface of the basin. Both formulas work for rectangular and circular basins, and neither formula depends on basin depth.

Calculate the weir loading in a plant that is operating at 8 mgd and has four sedimentation basins, each with a 200-ft weir. 8,000,000/(4 × 200 ft) = 10,000 gpd/ft.

Calculate the surface overflow rate for the same plant if each basin is 38 ft × 106 ft. 8,000,000/(4 × 38 ft × 106 ft) = approximately 500 gpd/ft^2.

Filter Loading and Backwash Rate

A WTP operates four filters, each of which is 32 ft × 26 ft. What is the average filtration rate if the plant flow is 8 mgd? 8,000,000 gpd/1,440 = 5,555 gpm; (4)(32 ft × 26 ft) = 3,328 ft^2; 5,555 gpm/3,328 ft^2 = 1.67 gpm/ft^2.

An operator needs to calculate the average wash time for a filter that measures 14 ft × 26 ft. If the filter was washed 46 times, and 2,511,600 gal was used during that span of time at an average backwash rate of 15 gpm/ft^2, what was the average length of backwash? Area of filter is 14 ft × 26 ft = 364 ft^2; 2,511,600 gal/46 washes = 54,600 gal/wash; 15 gpm/ft^2 × 364 ft^2 = 5,460 gpm during the wash; (54,600 gal/wash)/(5460 gpm) = 10 min/wash, on average.

Horsepower and Pumping

The power required to drive a pump is called water horsepower (whp). Because pumps are inefficient, the power requirements for the pump are calculated as brake horsepower (bhp). The motors that drive pumps are also inefficient; therefore, that power requirement calculation is called motor horsepower (mhp) (see Figure B-3).

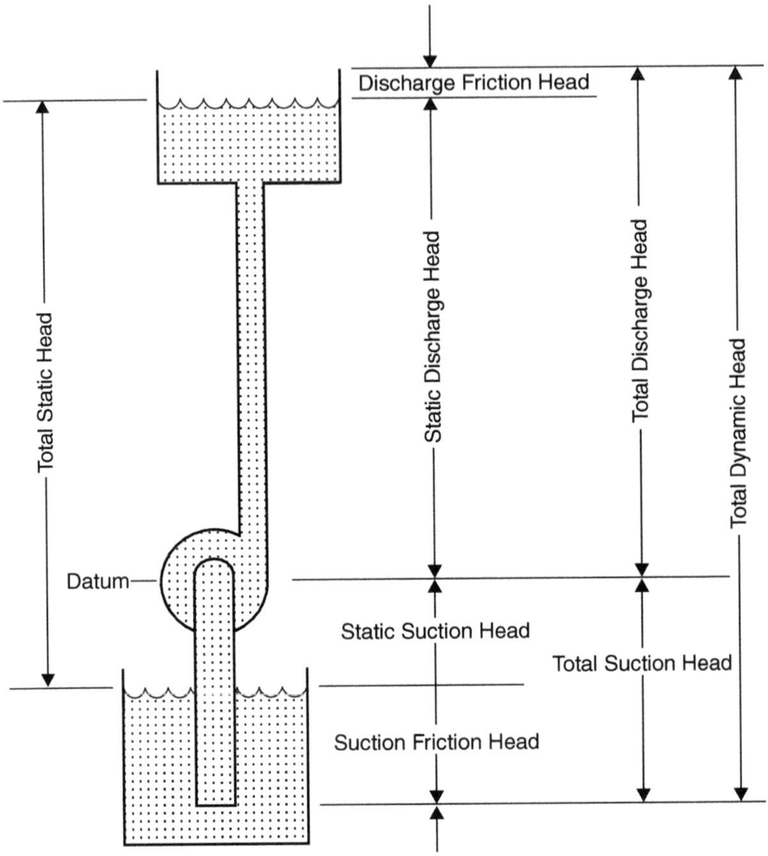

Figure B-3 Schematic illustrating common pumping terminology

What is the whp requirement to pump 375 gpm against a head of 55 psi? Formula for whp = (gpm × head in feet)/3,960. Therefore, (375 × (55 × 2.31))/3,960 = 12 hp.

If the pump in the previous problem is 87% efficient, what is the bhp requirement? bhp = whp/Efficiency = 12/0.87 = 13.8 bhp.

If the motor that drives the pump in the previous problems is 92% efficient, what is the mhp requirement? mhp = bhp/Efficiency = 13.8/0.92 = 15 mhp.

The previous examples show that it always takes more energy to move water through systems because inefficiencies result from friction and heat loss.

What is the whp required to fill an elevated storage tank with a pump operating at 5,600 gpm if the static head is 85 psi and the suction head is 9 ft? 85 psi × 2.31 = 196.35 ft; 196.35 ft – 9 ft = 187.35 ft; therefore, (5,600 gpm × 187.35)/3,960 = 265 hp.

Concentration Times Time ($C \times T$)

Using Figure B-4 and Table B-1, calculate the reportable $C \times T$ value in milligram minutes per liter. The highest source water flow and lowest clearwell depth were at 1,800 hours. Therefore, $C \times T$ = (44 ft × 30 ft × 8.3 ft water depth)(7.48)(0.34 baffling factor)(1.6 mg/L)/300 gpm = 148.6 mg-min/L.

Calculate the contact time (CT) for a clearwell contact basin with a baffling factor of 0.5, a capacity of 2.4 mil gal, plant flow of 5 mgd, and a chlorine residual at the basin outlet of 0.9 mg/L. Detention time = Volume/Flow, so T = 2.4 mg/5 mgd = 0.48 days × 1440 min/d = 691.2 min. Now multiply by 0.5 baffling factor, so the allowed detention time is 345.6 min. The CT is calculated by multiplying this by the residual. CT = 0.9 mg/L × 345.6 min = 311 min-mg/L.

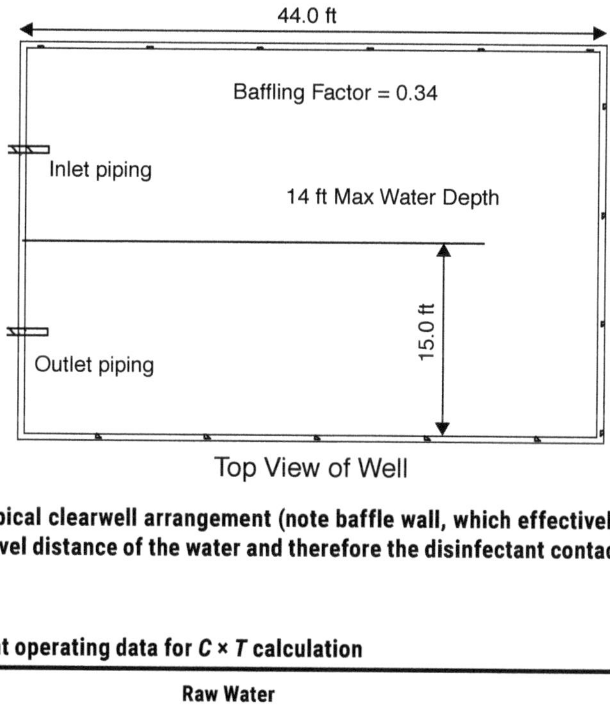

Figure B-4 Typical clearwell arrangement (note baffle wall, which effectively increases the travel distance of the water and therefore the disinfectant contact time)

Table B-1 Plant operating data for $C \times T$ calculation

Time	Clearwell Depth	Cl_2 Residual	Raw Water Flow, gpm	Time	Clearwell Depth	Cl_2 Residual	Raw Water Flow, gpm
0000	11.0	1.5	220	1200	9.6	1.5	175
0100	11.0	1.5	220	1300	9.0	1.5	175
0200	10.9	1.5	220	1400	8.7	1.5	175
0300	10.9	1.5	220	1500	8.5	1.5	220
0400	10.6	1.4	220	1600	8.4	1.6	220
0500	10.5	1.5	220	1700	8.3	1.6	220

(continued)

Table B-1 Plant operating data for *C* × *T* calculation (*Continued*)

Time	Clearwell Depth	Cl_2 Residual	Raw Water Flow, gpm	Time	Clearwell Depth	Cl_2 Residual	Raw Water Flow, gpm
0600	10.0	1.5	280	1800	8.3	1.6	300
0700	9.8	1.7	270	1900	8.6	1.6	280
0800	9.7	1.7	270	2000	9.0	1.5	280
0900	9.7	1.7	270	2100	9.2	1.5	220
1000	9.9	1.6	220	2200	9.5	1.5	220
1100	10.0	1.6	230	2300	9.9	1.5	220

Lime Softening

Total hardness (milligrams per liter of calcium carbonate [$CaCO_3$]) = carbonate hardness (milligrams per liter of $CaCO_3$) + noncarbonate hardness (milligrams per liter of $CaCO_3$), carbonate hardness = total alkalinity. What is the noncarbonated hardness if the alkalinity is 120 mg/L and total hardness is 147 mg/L? Total hardness – total alkalinity = noncarbonated hardness. 147 – 120 = 27 mg/L noncarbonated hardness.

Calculating the chemical dosages for the excess lime softening process can be involved. These calculations are too lengthy to present in this handbook. Consult the book *Math for Water Treatment Operators: Practice Problems to Prepare for Water Treatment Operator Certification Exams* (Giorgi 2007) in the bibliography section for detailed step-by-step guidance for these calculations.

Floor Loading

The floor loading capacity in the chemical storage room is 438 lb/ft². How high can you stack 100-lb bags of soda ash if each bag is 30 in. × 18 in. × 6 in.? Each bag is 2.5 ft × 1.5 ft × 0.5 ft. Surface area of one bag is 2.5 ft × 1.5 ft = 3.75 ft². Therefore, 3.75 ft² × 438 lb/ft² = 1,642.5 lb, and 1,642.5 lb × 1 bag/100 lb = 16.4 bags. 16.4 bags × 0.5 ft/bag = 8-ft maximum height.

BIBLIOGRAPHY

Giorgi, J. 2010 (6th ed.). *Water Operator Certification Study Guide.* Denver: AWWA.

Giorgi, J. 2007. *Math for Water Treatment Operators: Practice Problems to Prepare for Water Treatment Operator Certification Exams.* Denver: AWWA.

Appendix C

Water Chemistry

All matter exists in one or more of three states: solid, liquid, or gas. Some substances, such as water, can exist in all three. Simply adding or taking away heat can often bring about the transition from one state to another. Water is a prime example. If we add heat to ice, it will eventually change to a liquid state. This is a physical property of water. We can reverse this change simply by taking away enough heat to freeze the water.

Not all substances will behave similarly. Add heat to a solid piece of wood and it will not normally liquefy—it will char and then burn. This describes a chemical reaction. Once this change has taken place, we cannot go back to the piece of wood. The change was irreversible. The composition and properties of the wood have changed. The study of these types of changes and the nature of compositions of materials is called chemistry.

All of the solid, liquid, and gaseous matter that we know of is composed of one or more elements. There are more than 100 known elements; all of them are listed in the periodic table of the elements. Table C-1 lists some of the more common elements and their properties. For most purposes, it is best to think of elements as the basic form of matter—those that cannot be broken down further into smaller components by ordinary means. When these elements combine, they form compounds. For example, water is a compound formed from the elements of hydrogen and oxygen. Compounds can usually be broken down to their individual elements.

NOTATION AND FORMULA FORMATS

Every element is represented by a symbol, which is a kind of shorthand to make it easier to read and display chemical formulas. Chemical formulas tell us the number of atoms of each element that make up a compound. For example, the symbol for hydrogen is H, and the symbol for oxygen is O. When put together in the ratio of two atoms of hydrogen to one atom of oxygen (H_2O), they represent the formula for water. The formula for carbon dioxide is CO_2 and shows that the compound is made up of one carbon atom and two oxygen atoms.

Table C-1 Chemical names and symbols for some elements common to water treatment

Element	Chemical Symbol	Atomic Weight	Atomic Number	Valence
Aluminum	Al	26.98	13	3
Calcium	Ca	40.08	20	2
Carbon	C	12.0	6	2,4
Chlorine	Cl	35.45	17	1,3,5,7
Copper	Cu	63.54	29	1,2
Fluorine	F	19	9	1
Hydrogen	H	1.008	1	1
Iron	Fe	55.85	26	2,3
Magnesium	Mg	24.31	12	2
Manganese	Mn	54.99	25	2,3,4,6,7
Nitrogen	N	14.01	7	3,5
Oxygen	O	16	8	2
Phosphorus	P	30.98	15	3,5
Potassium	K	39.1	19	1
Sodium	Na	22.99	11	1
Sulfur	S	32.06	16	2,4,6

ATOMIC WEIGHT AND MOLECULAR WEIGHT

Each element has an atomic weight, or formula weight, that helps to further determine the properties of that element. Referring to Table C-1, we see that the atomic weight of hydrogen is 1, and the atomic weight of oxygen is 16. When compounds form, the formula for the compound can be used to determine the molecular weight of the compound. In the example for water, two hydrogens and one oxygen combine to form a compound that has a molecular weight of 18. The percentage of weight that an element contributes to the overall molecule is determined by dividing the elemental weight by the total molecular weight. The percentage of hydrogen in the water molecule is $^{2}/_{18}$, or about 11%.

A molecule is the smallest unit of a compound. It is usually made up of several elements in combination but can be made up of two or more of the same elements. The aforementioned example shows a molecule of water being made up of two elements. Chlorine, Cl, is an element, but it usually exists as Cl_2, the compound called chlorine, which is a molecule. Using Table C-1, we see that the element Cl has a weight of about 35.5, but the chlorine molecule has an atomic weight of 70.

CHEMICAL LAWS, FORMULAS, CALCULATIONS, AND EQUATIONS

Chemistry has laws—i.e., predictable results for situations that never change. These laws can help us to make assumptions whenever we work with chemicals in the plant. One law, the Law of Conservation of Matter, states that matter can be neither created nor destroyed, only changed from one form to another. For example, if you bring 1 oz of water to a boil, it eventually turns to steam. The same amount of water is still available, just in a different form.

Other laws of chemistry, such as the Law of Definite Proportions and the Law of Multiple Proportions, state that "if a given compound is formed by the combination of elements, the proportion, by weight, of the element or elements making up the compound is always the same regardless of the method used to form the compound." Also, if "two or more compounds are composed of the same elements, the different weights of one element which combine with a fixed weight of another stand to each other in the ratio of small whole numbers." For example, water (H_2O) and hydrogen peroxide (H_2O_2) are made up of the same elements. Each compound has two hydrogens, but the number of oxygens in the peroxide is twice that of the number found in water. Therefore, the weight of oxygen found in the peroxide is exactly twice the amount found in water.

The Law of Combining Weight states that in every compound, the proportions, by weight, of each element may be expressed by a definite number (atomic weight) or an integral multiple of that number. This is an extension of the Law of Multiple Proportions.

These laws can be used to construct chemical equations, which are used to perform mathematical calculations that predict reactions. Chemical equations are simple expressions that consist of chemical formulas that indicate chemical changes, or reactions, that take place when we bring compounds together. We can show, for example, what happens when chlorine gas is put into water. The equation is as follows:

$$Cl_2 + H_2O \rightarrow HOCl + HCl$$

and tells us that a reaction took place and that reaction produced hypochlorous acid (HOCl) and hydrochloric acid (HCl). It also tells us that it takes one molecule of each of the reactants to produce one molecule of each of the products. We do not sometimes get two molecules of HOCl—we always get one. We also see that matter has been conserved: There are two chlorines on the left side of the equation and two on the right. Hydrogen and oxygen have also been conserved, although changed.

Sometimes the examples are not so simple—they must be further refined so that none of the laws of chemistry are violated. For example, we know that iron, if exposed to moisture, will oxidize into rust, which is Fe_2O_3. Iron is Fe, and oxygen, like chlorine, exists as a molecule consisting of two elements and is written as O_2. Therefore, if we combine the two compounds that make rust—i.e., iron and oxygen—we should get a formula that works. But when we do, it looks like this:

$$Fe + O_2 \rightarrow Fe_2O_3$$

We know that this must be wrong because there is one iron on the left side of the equation but two on the right. Similarly, there are two oxygens on the left and three on the right. Because we cannot create matter, we have to refine, or balance, the equation in order for it to make sense. Balancing equations requires that we start with the correct formulas for the compounds and that we know all of the compounds involved in the reaction. It may seem like a hit-or-miss experience, but, with practice, the simple equations can be balanced by anyone. (More complex equations exist, but they are not covered in this text.) In this example, we might start by putting a 2 in front of the Fe_2O_3. Now our formula looks like this:

$$Fe + O_2 \rightarrow 2Fe_2O_3$$

We now have six oxygens on the right side of the equation, but we need six on the left. If we put a 3 in front of the O_2, we would have six there, and the equation would read as follows:

$$Fe + 3O_2 \rightarrow 2Fe_2O_3$$

We are now balanced for oxygen. However, there's still more work to do. The Fe portion of the equation is out of balance—there are four on the right and only one on the left. If we put a 4 in front of the Fe, the equation becomes this:

$$4Fe + 3O_2 \rightarrow 2Fe_2O_3$$

The equation is now balanced; it took four molecules of iron to combine with three molecules of oxygen to form two molecules of rust.

This information can be put to practical use, predicting the amounts of chemicals needed to produce a desired amount of product. In the example of rusting of iron by oxygen, we saw the proportions of compounds needed to make the product called rust. Using Table C-1, we can determine that the atomic weight of iron is 55.85, and the atomic weight of oxygen is 16. Now we can give an account of the weight units of all of the atoms present as follows:

$$\text{Four atoms of iron} \rightarrow 4 \times 55.85 = 223.4$$

The O_2 atom is $2 \times 16 = 32$, and there are three of them, so $3 \times 32 = 96$. The iron oxide, or rust, is equivalent to 2×55.85 (two irons) plus 3×16 (three oxygens), or 159.7. Because there are two molecules of rust produced, the overall equivalent for it is 2×159.7, or 319.4. We have learned that 223.4 weight units of iron will combine with 96 weight units of oxygen to form 319.4 weight units of rust. The weight units can be anything—grams, pounds, tons, or ounces—as long as they are the same for each molecule. So, we can say that it takes 223.4 tons of iron and 96 tons of oxygen to form 319.4 tons of rust, and no other proportions will do—it always takes these amounts.

Appendix D

Metric Conversions

LINEAR MEASUREMENT

inch (in.)	× 25.4	= millimeters (mm)
inch (in.)	× 2.54	= centimeters (cm)
foot (ft)	× 304.8	= millimeters (mm)
foot (ft)	× 30.48	= centimeters (cm)
foot (ft)	× 0.3048	= meters (m)
yard (yd)	× 0.9144	= meters (m)
mile (mi)	× 1,609.3	= meters (m)
mile (mi)	× 1.6093	= kilometers (km)
millimeter (mm)	× 0.03937	= inches (in.)
centimeter (cm)	× 0.3937	= inches (in.)
meter (m)	× 39.3701	= inches (in.)
meter (m)	× 3.2808	= feet (ft)
meter (m)	× 1.0936	= yards (yd)
kilometer (km)	× 0.6214	= miles (mi)

AREA MEASUREMENT

square meter (m^2)	× 10,000	= square centimeters (cm^2)
hectare (ha)	× 10,000	= square meters (m^2)
square inch ($in.^2$)	× 6.4516	= square centimeters (cm^2)
square foot (ft^2)	× 0.092903	= square meters (m^2)
square yard (yd^2)	× 0.8361	= square meters (m^2)
acre	× 0.004047	= square kilometers (km^2)
acre	× 0.4047	= hectares (ha)
square mile (mi^2)	× 2.59	= square kilometers (km^2)
square centimeter (cm^2)	× 0.16	= square inches ($in.^2$)
square meters (m^2)	× 10.7639	= square feet (ft^2)
square meters (m^2)	× 1.1960	= square yards (yd^2)
hectare (ha)	× 2.471	= acres

square kilometer (km^2)	× 247.1054	= acres
square kilometer (km^2)	× 0.3861	= square miles (mi^2)

VOLUME MEASUREMENT

cubic inch ($in.^3$)	× 16.3871	= cubic centimeters (cm^3)
cubic foot (ft^3)	× 28,317	= cubic centimeters cm^3)
cubic foot (ft^3)	× 0.028317	= cubic meters (m^3)
cubic foot (ft^3)	× 28.317	= liters (L)
cubic yard (yd^3)	× 0.7646	= cubic meters (m^3)
acre foot (acre-ft)	× 1233.48	= cubic meters (m^3)
ounce (US fluid) (oz)	× 0.029573	= liters (L)
quart (liquid) (qt)	× 946.9	= milliliters (mL)
quart (liquid) (qt)	× 0.9463	= liters (L)
gallon (gal)	× 3.7854	= liters (L)
gallon (gal)	× 0.0037854	= cubic meters (m^3)
peck (pk)	× 0.881	= decaliters (dL)
bushel (bu)	× 0.3524	= hectoliters (hL)
cubic centimeters (cm^3)	× 0.061	= cubic inches ($in.^3$)
cubic meter (m^3)	× 35.3183	= cubic feet (ft^3)
cubic meter (m^3)	× 1.3079	= cubic yards (yd^3)
cubic meter (m^3)	× 264.2	= gallons (gal)
cubic meter (m^3)	× 0.000811	= acre-feet (acre-ft)
liter (L)	× 1.0567	= quart (liquid) (qt)
liter (L)	× 0.264	= gallons (gal)
liter (L)	× 0.0353	= cubic feet (ft^3)
decaliter (dL)	× 2.6417	= gallons (gal)
decaliter (dL)	× 1.135	= pecks (pk)
hectoliter (hL)	× 3.531	= cubic feet (ft^3)
hectoliter (hL)	× 2.84	= bushels (bu)
hectoliter (hL)	× 0.131	= cubic yards (yd^3)
hectoliter (hL)	× 26.42	= gallons (gal)

(Note: US gallons are listed above.)

PRESSURE MEASUREMENT

pounds/square inch (psi)	× 6.8948	= kilopascals (kPa)
pounds/square inch (psi)	× 0.00689	= pascals (Pa)
pounds/square inch (psi)	× 0.070307	= kilograms/square centimeter (kg/cm^2)
pounds/square foot (lb/ft^2)	× 47.8803	= pascals (Pa)
pounds/square foot (lb/ft^2)	× 0.000488	= kilograms/square centimeter (kg/cm^2)
pounds/square foot (lb/ft^2)	× 4.8824	= kilograms/square meter (kg/m^2)
inches of mercury	× 3376.8	= pascals (Pa)
inches of water	× 248.84	= pascals (Pa)
bar	× 100,000	= newtons per square meter (N/m^2)

pascals (Pa)	×	1	=	newtons per square meter (N/m^2)
pascal (Pa)	×	0.000145	=	pounds/square inch (psi)
kilopascals (kPa)	×	0.145	=	pounds/square inch (psi)
pascal (Pa)	×	0.000296	=	inches of mercury (at 60°F)
kilogram/square centimeter (k/cm^2)	×	14.22	=	pounds/square inch (psi)
kilogram/square centimeter (k/cm^2)	×	28.959	=	inches of mercury (at 60°F)
kilogram/square meter (k/m^2)	×	0.2048	=	pounds per square foot (lb/ft^2)
centimeters of mercury	×	0.4461	=	feet of water

WEIGHT MEASUREMENT

ounce (oz)	×	28.3495	=	grams (g)
pound (lb)	×	0.045359	=	grams (g)
pound (lb)	×	0.4536	=	kilograms (kg)
ton (short)	×	0.9072	=	megagrams (metric ton)
pounds/cubic foot (lb/ft^3)	×	16.02	=	grams per liter (g/L)
pounds/million gallons (lb/mil gal)	×	0.1198	=	grams per cubic meter (g/m^3)
gram (g)	×	15.4324	=	grains (gr)
gram (g)	×	0.0353	=	ounces (oz)
gram (g)	×	0.0022	=	pounds (lb)
kilograms (kg)	×	2.2046	=	pounds (lb)
kilograms (kg)	×	0.0011	=	tons (short)
megagram (metric ton)	×	1.1023	=	tons (short)
grams/liter (g/L)	×	0.0624	=	pounds per cubic foot (lb/ft^3)
grams/cubic meter (g/m^3)	×	8.3454	=	pounds/million gallons (lb/mil gal)

FLOW RATES

gallons/second (gps)	×	3.785	=	liters per second (L/s)
gallons/minute (gpm)	×	0.00006308	=	cubic meters per second (m^3/s)
gallons/minute (gpm)	×	0.06308	=	liters per second (L/s)
gallons/hour (gph)	×	0.003785	=	cubic meters per hour (m^3/h)
gallons/day (gpd)	×	0.000003785	=	million liters per day (ML/d)
gallons/day (gpd)	×	0.003785	=	cubic meters per day (m^3/d)
cubic feet/second (ft^3/s)	×	0.028317	=	cubic meters per second (m^3/s)
cubic feet/second (ft^3/s)	×	1,699	=	liters per minute (L/min)
cubic feet/minute (ft^3/min)	×	472	=	cubic centimeters per second (cm^3/s)
cubic feet/minute (ft^3/min)	×	0.472	=	liters per second (L/s)
cubic feet/minute (ft^3/min)	×	1.6990	=	cubic meters per hour (m^3/h)
million gallons/day (mgd)	×	43.8126	=	liters per second (L/s)
million gallons/day (mgd)	×	0.003785	=	cubic meters per day (m^3/d)
million gallons/day (mgd)	×	0.043813	=	cubic meters per second (m^3/s)
gallons/square foot (gal/ft^2)	×	40.74	=	liters per square meter (L/m^2)
gallons/acre/day (gal/acre/d)	×	0.0094	=	cubic meters/hectare/day ($m^3/ha/d$)

gallons/square foot/day ($gal/ft^2/d$)	× 0.0407	= cubic meters/square meter/day ($m^3/m^2/d$)
gallons/square foot/day ($gal/ft^2/d$)	× 0.0283	= liters/square meter/day ($L/m^2/d$)
gallons/square foot/minute ($gal/ft^2/min$)	× 2.444	= cubic meters/square meter/hour ($m^3/m^2/h$) = m/h
gallons/square foot/minute ($gal/ft^2/min$)	× 0.679	= liters/square meter/second ($L/m^2/s$)
gallons/square foot/minute ($gal/ft^2/min$)	× 40.7458	= liters/square meter/minute ($L/m^2/min$)
gallons/capita/day (gpcd)	× 3.785	= liters/day/capita (L/d per capita)
liters/second (L/s)	× 22,824.5	gallons per day (gpd)
liters/second (L/s)	× 0.0228	= million gallons per day (mgd)
liters/second (L/s)	× 15.8508	= gallons per minute (gpm)
liters/second (L/s)	× 2.119	= cubic feet per minute (ft^3/min)
liters/minute (L/min)	× 0.0005886	= cubic feet per second (ft^3/s)
cubic centimeters/second (cm^3/s)	× 0.0021	= cubic feet per minute (ft^3/min)
cubic meters/second (m^3/s)	× 35.3147	= cubic feet per second (ft^3/s)
cubic meters/second (m^3/s)	× 22.8245	= million gallons per day (mgd)
cubic meters/second (m^3/s)	× 15,850.3	= gallons per minute (gpm)
cubic meters/hour (m^3/s)	× 0.5886	= cubic feet per minute (ft^3/min)
cubic meters/hour (m^3/s)	× 4.403	= gallons per minute (gpm)
cubic meters/day (m^3/d)	× 264.1720	= gallons per day (gpd)
cubic meters/day (m^3/d)	× 0.00026417	= million gallons per day (mgd)
cubic meters/hectare/day ($m^3/ha/d$)	× 106.9064	= gallons per acre per day (gal/acre/d)
cubic meters/sq meter/day ($m^3/m^2/d$)	× 24.5424	= gallons/square foot/day ($gal/ft^2/d$)
liters/square meter/minute ($L/m^2/min$)	× 0.0245	= gallons/square foot/minute ($gal/ft^2/min$)
liters/square meter/minute ($L/m^2/min$)	× 35.3420	= gallons/square foot/day ($gal/ft^2/d$)

WORK, HEAT, AND ENERGY

British thermal units (Btu)	× 1.0551	= kilojoules (kJ)
British thermal units (Btu)	× 0.2520	= kilogram-calories (kg-cal)
foot-pound (force) (ft-lb)	× 1.3558	= joules (J)
horsepower-hour (hp·h)	× 2.6845	= megajoules (MJ)
watt-second (W·s)	× 1.000	= joules (J)
watt-hour (W·h)	× 3.600	= kilojoules (kJ)
kilowatt-hour (kW·h)	× 3,600	= kilojoules (kJ)
kilowatt-hour (kW·h)	× 3,600,000	= joules (J)
British thermal units per pound (Btu/lb)	× 0.5555	= kilogram-calories per kilogram (kg-cal/kg)
British thermal units per cubic foot (Btu/ft^3)	× 8.8987	= kilogram-calories/cubic meter ($kg\text{-}cal/m^3$)

kilojoule (kJ)	×	0.9478	=	British thermal units (Btu)
kilojoule (kJ)	×	0.00027778	=	kilowatt-hours (kW·h)
kilojoule (kJ)	×	0.2778	=	watt-hours (W·h)
joule (J)	×	0.7376	=	foot-pounds (ft-lb)
joule (J)	×	1.0000	=	watt-seconds (W·s)
joule (J)	×	0.2399	=	calories (cal)
megajoule (MJ)	×	0.3725	=	horsepower-hour (hp·h)
kilogram-calories (kg-cal)	×	3.9685	=	British thermal units (Btu)
kilogram-calories per kilogram (kg-cal/kg)	×	1.8000	=	British thermal units per pound (Btu/lb)
kilogram-calories per liter (kg-cal/L)	×	112.37	=	British thermal units per cubic foot (Btu/ft^3)
kilogram-calories/cubic meter (kg-cal/m^3)	×	0.1124	=	British thermal units per cubic foot (Btu/ft^3)

VELOCITY, ACCELERATION, AND FORCE

feet per minute (ft/min)	×	18.2880	=	meters per hour (m/h)
feet per hour (ft/h)	×	0.3048	=	meters per hour (m/h)
miles per hour (mph)	×	44.7	=	centimeters per second (cm/s)
miles per hour (mph)	×	26.82	=	meters per minute (m/min)
miles per hour (mph)	×	1.609	=	kilometers per hour (km/h)
feet/second/second (ft/s^2)	×	0.3048	=	meters/second/second (m/s^2)
inches/second/second (in./s^2)	×	0.0254	=	meters/second/second (m/s^2)
pound force (lbf)	×	4.44482	=	newtons (N)
centimeters/second (cm/s)	×	0.0224	=	miles per hour (mph)
meters/second (m/s)	×	3.2808	=	feet per second (ft/s)
meters/minute (m/min)	×	0.0373	=	miles per hour (mph)
meters per hour (m/h)	×	0.0547	=	feet per minute (ft/min)
meters per hour (m/h)	×	3.2808	=	feet per hour (ft/h)
kilometers/second (km/s)	×	2.2369	=	miles per hour (mph)
kilometers/hour (km/h)	×	0.0103	=	miles per hour (mph)
meters/second/second (m/s^2)	×	3.2808	=	feet/second/second (ft/s^2)
meters/second/second (m/s^2)	×	39.3701	=	inches/second/second (in./s^2)
newtons (N)	×	0.2248	=	pounds force (lbf)

Author Biography

Nicholas G. Pizzi has worked in the drinking water industry since 1973 and has spent most of that time operating water treatment plants and helping others do so. Through American Water Works Association (AWWA), he has published books, courses, webinars, manuals, and scores of PowerPoint presentations that have been crafted to assist operators as they pursue licensing examinations. His belief in the role of the operators as the most important group of individuals in the drinking water field is evident throughout his work.

Along with performing tasks as a self-employed consultant since 1987, Nick served as vice-president of Environmental Engineering & Technology, Inc. (now Cornwell Engineering Group), an environmental engineering firm. Previously, he was the assistant commissioner of operations and water quality manager for the Cleveland, Ohio, Division of Water and director of water at the Lake County, Ohio, Water Division. He began his career in his hometown of Ashtabula, Ohio, as the chemist and water quality manager of the Ohio-American Water Company (now owned by Aqua Ohio).

Mr. Pizzi holds a Class IV water supply license in Ohio and a BA degree in chemistry from St. Vincent College in Latrobe, Pa. He was chair of the Partnership for Safe Water's Program Effectiveness Assessment Committee and a trustee for the Distribution and Plant Operations Division of AWWA. He acted as liaison to several AWWA committees and has served on various Water Research Foundation project advisory committees and as a *Journal AWWA* reviewer. He also served on the *Opflow* editorial advisory board.

He has been published in *Journal AWWA* and in AWWA's popular magazine for operators, *Opflow*. He has presented at AWWA annual conferences and AWWA's Water Quality Technology Conference. He served three years on the Technical and Educational Council and six years on the Standards Council. AWWA placed him on the Flint, Mich., Advisory Board to help with the mitigation work needed there, which led to operator training services in Flint at the request of the state of Michigan.

Over his distinguished career in the water sector, AWWA awarded Mr. Pizzi the *Opflow* Publication Award, Operator Meritorious Service Award, the Ohio Section Education Award, and the George Warren Fuller Award.

Index

Note: *f.* indicates figure; *t.* indicates table

C

I

J

K

L

M

Q

R

S

T

www.ingramcontent.com/pod-product-compliance
Ingram Content Group UK Ltd.
Pitfield, Milton Keynes, MK11 3LW, UK
UKHW020658041026
15590UKWH00001B/349

9 781647 172473